PRENTICE-HALL

FOUNDATIONS OF DEVELOPMENTAL BIOLOGY SERIES

Clement L. Markert, Editor

*Volumes published or in preparation:*

FERTILIZATION  *C. R. Austin*

CONTROL MECHANISMS IN PLANT DEVELOPMENT
*Arthur W. Galston and Peter J. Davies*

PRINCIPLES OF MAMMALIAN AGING
*Robert R. Kohn*

EMBRYONIC DIFFERENTIATION
*H. E. Lehman*

DEVELOPMENTAL GENETICS*  *Clement L. Markert and Heinrich Ursprung*

CELL REPRODUCTION DURING DEVELOPMENT
*David M. Prescott*

PATTERNS IN PLANT DEVELOPMENT
*T. A. Steeves and I. M. Sussex*

CELLS INTO ORGANS: The Forces That Shape
the Embryo  *J. P. Trinkaus*

* Published jointly in Prentice-Hall's *Foundations of Modern Genetics Series*

# DEVELOPMENTAL
# GENETICS

## Clement L. Markert
Yale University

## Heinrich Ursprung
Laboratory for Developmental Biology
Zurich, Switzerland

PRENTICE-HALL, INC.  Englewood Cliffs, New Jersey

We dedicate this book to
MARGARET MARKERT
and
URSULA URSPRUNG
who helped the writing in
innumerable ways

FOUNDATIONS OF DEVELOPMENTAL BIOLOGY SERIES

PRENTICE-HALL INTERNATIONAL, INC., London
PRENTICE-HALL OF AUSTRALIA, PTY. LTD., Sydney
PRENTICE-HALL OF CANADA, LTD., Toronto
PRENTICE-HALL OF INDIA PRIVATE LIMITED, New Delhi
PRENTICE-HALL OF JAPAN, INC., Tokyo

# Contents

# Foundations of DEVELOPMENTAL BIOLOGY

The development of organisms is so wondrous and yet so common that it has compelled man's attention and aroused his curiosity from earliest times. But developmental processes have proved to be complex and difficult to understand, and little progress was made for hundreds of years. By the beginning of this century, increasingly skillful experimentation began to accelerate the slow advance in our understanding of development. Most important in recent years has been the rapid progress in the related disciplines of biochemistry and genetics— progress that has now made possible an experimental attack on developmental problems at the molecular level. Many old and intractable problems are taking on a fresh appeal, and a tense expectancy pervades the biological community. Rapid advances are surely imminent.

New insights into the structure and function of cells are moving the principal problems of developmental biology into the center of scientific attention, and increasing numbers of biologists are focusing their research efforts on these problems. Moreover, new tools and experimental designs are now available to help in their solution.

At this critical stage of scientific development a fresh assessment is needed. This series of books in developmental biology is designed to provide essential background material and then to examine the frontier where significant advances are occurring or expected. Each book is written by a leading investigator actively concerned with the problems and concepts he discusses. Students at intermediate and advanced levels of preparation and investigators in other areas of biology should find these books informative, stimulating, and useful. Collectively, they present an authoritative and penetrating analysis of the major problems and concepts of developmental biology, together with a critical appraisal of the experimental tools and designs that make developmental biology so exciting and challenging today.

CLEMENT L. MARKERT

# Preface

The title of this book expresses our intention to view development in genetic terms. This was a common view at one time but then genetics and development went their separate ways, genetics to define the gene, its structure, function, mutation, and replication, and development to describe and analyze the interactions of cells and tissues. Genetics succeeded and was rightly crowned queen of the biological sciences. Development has continued to promise much but has so far produced little. The basic problems of development can be formulated in genetic terms and such formulations also express the central current problem of genetics—namely the regulation of gene function in metazoans. Thus, these two vast areas of biology meet on a common concern. Differential gene function—the origin of many cell phenotypes from a single genotype—that is the central contemporary problem of both genetics and development, and it is also the thesis of this book.

We hope that our book will help tie together the Prentice-Hall Foundations of Modern Genetics Series and the Developmental Biology Series. We have tried to be brief and sought to describe representative experiments or present data that illustrate the main features of the genetic foundations for cellular differentiation at several levels of cell organization. We have also treated several special topics such as pigmentation and neoplasia, for their own sake and because they seemed to illustrate significant aspects of developmental genetics. Several topics which the reader might expect to find have been omitted—the development of the immune system, e.g. This topic is developing so rapidly that any treatment seemed destined to be obsolete before it could appear in print. Certain topics such as metamorphosis and regeneration deserve book-length treatments. Others, such as pleiotropy, have been treated well enough by others, and we could add little more. Selected references at the end of each chapter should help the interested reader to gain access to more extensive treatments of each topic and perhaps to discover some that we have neglected. No attempt has been made to document our statements or to provide detailed citations to the original literature. This is not a research monograph, but rather a brief introduction to major areas of biological interest. Our audience should include the initiated undergraduate as well as the beginning graduate student. We believe they will all profit from a thoughtful perusal of this book. Many students and colleagues have helped us; we are grateful to them. We hope that the usefulness of this book will in part repay their interest and patience during the long gestation.

CLEMENT L. MARKERT
HEINRICH URSPRUNG

# ONE

## Differential Gene Function: The Basis of Cell Differentiation

How does a single cell—the fertilized egg—give rise to the many different kinds of cells that make up the bodies of metazoan organisms? This is the fundamental problem of developmental biology. Phrased in genetic terms it becomes the central question of developmental genetics: How does one cellular genotype give rise to many hundreds of different cellular phenotypes? The development of an embryo from a fertilized egg proceeds through a sequence of complex changing populations of cells; the characteristics of the individual cells and their organization into tissues and organs determine the properties of the embryo at each stage in its development. The fundamental unit of development, as of life itself, is the cell, and perhaps as many as a thousand distinct cell types arise from a single egg during the development of a complex organism such as a mammal.

During the last century, an important hypothesis pictured the egg as containing determinants from each cell type that could arise during the course of embryonic development. These hypothetical determinants were assumed to be parceled out during cell division so that each cell received its characteristic quota, and thus its properties were specified and fixed. Such a mechanism for cell differentiation was plausible and seemed adequate to account for the origin of different cell types, even though a very precise and orderly cleavage pattern was required. Rigidly fixed cleavage patterns were, in fact, observed in the embryos

1

of many kinds of invertebrates. Moreover, experimental analysis supported the determinant hypothesis since deletions of part of an embryo by microsurgery commonly lead to corresponding defects at later stages of development without greatly disturbing the remainder of the embryo. From these observations, embryos were thought to be mosaics constructed of independent cells, each of which developed without regard to adjacent cells but all of which were placed together in a precise pattern by a fixed program of cell division. This view of the origin of specific cell types through the distribution of hypothetical determinants was soon shown by experimentation to be unacceptable as a generalization applicable to all embryos. Many embryos were found to be highly regulative in their development and to be made up of blastomeres with equivalent potentialities that could be elicited by experimental rearrangements of the cells of the embryo.

Separation of the blastomeres at the two-cell stage of certain embryos —sea urchins, for example—provoked each blastomere to develop into an entire normal larva, although of half normal size. Without experimental intervention each blastomere would, of course, have given rise to only one-half an animal. At later stages of development, particularly in amphibian embryos at the blastula stage, it was possible surgically to interchange groups of cells and to observe their development in new tissue environments. The transplanted cells frequently differentiated in a way appropriate to their new location in the embryo, and by so doing demonstrated again that the fate of an embryonic cell is not fixed by any selective allocation of determinants during cell division but is, in fact, dependent upon a sequence of interactions with the tissue environment. In these regulative embryos, each cell carried within itself the potentiality for a wide variety of differentiated states, even though only one of these states would ultimately be selected for expression in the normal course of development.

With the realization that the potentialities of a cell are far greater than the actualities expressed during development, investigators were confronted for the first time with the clear problem of discovering what it is that selects among these numerous potentialities. One of the principal goals of current experimentation in the area of developmental genetics is to describe in molecular terms the nature of the mechanisms that differentiate cells by selectively calling forth a restricted fraction of their potentialities. Our present knowledge of genetics makes clear that the properties of a differentiated cell are an ultimate reflection of the activity of its genes. From cytological observations and from experiments to be described later, it seems probable that during cell division identical sets of chromosomes, and thus probably identical sets of genes, are distributed to each daughter cell. Yet through successive cell divisions the cells gradually diverge until they differentiate into cells of very different phenotypes—into muscle, nerve, pigment, or gland cells, for example. Although these cells are conspicuously different in phenotype, they presumably have exactly the same genotype. Obviously, each cell manifests only a small fraction of its genes. The cells do behave as if different sets of active genes (determinants) were parceled out to them during the repeated cell divisions of embryonic development. We now realize that the error of such a formulation lies in attributing the differences among the cells to a distribution mechanism that apportions the genes unequally. It is not the genes or the chromosomes that are distributed unequally but rather the chemical environments of the chromosomes—environments that are eventually expressed in the differential regulation of gene function.

The distinctions among differentiated cells can be attributed largely, if not entirely, to different enzymatic compositions. We know that the structure of these enzymes, as well as that of other proteins, is encoded in the sequence of nucleotides making up the deoxyribonucleic acid of the chromosomes, of the genes, and therefore we can conclude that the appearance of a specific protein in the cell indicates that the corresponding gene is active in some cell. Only a few proteins are known to be exchanged between cells—for example, protein hormones—and thus qualitative differences in protein content among cells nearly always indicate corresponding differences in the repertory of functioning genes in those cells. The genes for hemoglobin synthesis, for example, function early in the life of erythrocytes but not in other cells. The gene for tyrosinase synthesis functions only during late stages in the differentiation of the melanoblast and not in other cells. The gene for insulin synthesis is active only in the islet cells of the pancreas. Many such examples of time and cell specificity of protein appearance in cells can be cited. Not only is each cell characterized by a distinctive assortment of proteins but these are specific for each stage of its differentiation. Some proteins appear at early stages and then disappear— for example, fetal hemoglobin. Others, such as tyrosinase, appear only at terminal stages in cell differentiation. Thus, the fraction of the genome that is active and identifiable by the production of specific proteins is specific for each cell at each stage of its development. We are forced to the conclusion that differential gene function is an intrinsic and fundamental aspect of cell differentiation. Whatever is responsible for regulating gene function must also be responsible for cellular differentiation and, indirectly, for the morphological changes occurring during the course of embryonic development.

Although the terminal phenotype of a cell is clearly a consequence of the selective expression of its genes, this selective expression could be achieved at several steps in cellular metabolism. The relative activity among the genes might be altered by differential replication of certain genes so that the more numerous genes would contribute more. The ribonucleic acid molecules transcribed from the DNA might be selectively regulated within the nucleus through altering the relative rates of synthesis or by differential degradation of RNA. The passage of RNA from the nucleus into the cytoplasm might also be regulated so that only certain varieties of RNA would become available for affecting the phenotype of the cell. The functional activity of the RNA in the cytoplasm might be regulated by many metabolic devices to alter the final proportions among the protein products of the RNA; that is, translation of the RNA might be an accessible event for effectively regulating the expression of the genome. Once a protein has been synthesized, it may be modified in various ways to alter its structure and function. Such modification may be described as epigenetic. There is some evidence for the regulation of gene expression at nearly all these steps in cell metabolism. This evidence will be discussed in subsequent chapters.

There is a widespread conviction among biologists that the DNA of the cell, whether in chromosomes or in other organelles, is the sole source of information in the life of the cell, and therefore also in cellular differentiation. This view begins with the demonstration that the sequence of nucleotides in the DNA is responsible for the linear sequence of amino acids in proteins—for the primary structure. Extrapolating from this foundation, we assume that the functional three-dimensional structure of proteins is always a single intrinsic expression of the primary structure and is achieved without any additional informational input from the metabolic machinery of the cell. The proteins in their three-dimensional configurations then function as enzymes, or in other ways. They also are assumed to self-assemble into structures of higher order to form cell organelles, such as mitochondria, membranes, and so forth. These structures, together with the molecular products of their activity, make up the cell and are responsible for all its characteristics.

The primary role of DNA as a source of information for the construction of a cell in its several stages of differentiation seems well established. However, the cell is an exceedingly complicated structure composed of many different kinds of molecules and organelles. The arrangement of these materials in the cell—that is, the organization of these molecules into structures of higher order and into integrated patterns—suggests the possibility that at least two additional sources of information aside from the DNA may be present in the egg and important to the development of the embryo.

One source of "information" that is passed from cell to cell is in the form of highly organized, cyclical, chemical reaction systems, in which the product of one reaction step is the substrate for the next. The enzymatic production of amino acids, which, in turn, compose the enzymes, exemplifies such a cyclical series of reactions. No link in the chain of reactions could be broken without destroying the entire sequence. The requirement that glycogen be present before more can be synthesized is a similar example. Quite obviously, the molecular constituents of such metabolic reaction systems are transmitted through the egg to succeeding generations in a fashion that is effectively independent of immediate intervention by the chromosomal genes.

A second and perhaps more significant source of information lies in the organization of macromolecules into aggregates of a higher order, such as cell organelles, particularly membranes. The components of these structures are obviously arranged together in a manner consistent with their physicochemical properties. However, such structures might not arise quickly enough within the cell to fulfill the physiological requirements of the cell if the constituent molecules assembled only by random collisions. It seems possible that some of these macromolecular structures may act as templates for the rapid assembly of additional molecules to make more of the same kind of structure. Mitochondria are known to replicate; cell membranes are always present and can expand rapidly. If the preexisting structure of these organelles is of critical importance in the formation of more of the same, then molecular accidents in assembly might change the structure of such an organelle; the new "mutated" structure might be perpetuated if it proved advantageous to the cell. Such historical accidents of molecular aggregation would constitute a source of new information analogous to accidental, that is, mutational, changes in the nucleotide sequences of DNA.

## REFERENCES

*Including Extensive Introductions to Developmental Genetics*

Modern review or symposium articles on developmental genetics form excellent supplements to the contents of this book. Such articles may be found in: Advances in Genetics; Progress in Nucleic Acid Research and Molecular Biology; Annual Review of Biochemistry; Annual Review of Genetics; Cold Spring Harbor Symposia on Quantitative Biology; National Cancer Institute Monographs; Progress in Biophysics and Molecular Biology; Results and Problems in Cell Differentiation; Current Topics in Developmental Biology;

Experimental Biology and Medicine; Symposia of the Society for Developmental Biology (since 1967 as Supplements to the journal, Developmental Biology). The current literature on developmental genetics is scattered among many journals. Important articles on developmental genetics are to be found in: Genetics; Journal of Molecular Biology; Molecular and General Genetics (formerly Zeitschrift für Vererbungslehre); Biochemical Genetics; The Journal of Experimental Zoology; Chromosoma; Roux' Archiv; Developmental Biology; Journal of Cell Biology; Journal of Cell Science.

Bell, E. 1965. Molecular and cellular aspects of developmental biology. Harper and Row, New York. A collection containing many pacemaking articles in the field of developmental biology, by a large number of authors, with connecting comments by the editor.

Brown, D. D. 1969. Developmental genetics. *In* Annual review of genetics, vol. III. Annual Reviews, Inc., Palo Alto, Calif. An up-to-date and critical evaluation of developmental genetics, with emphasis on cell determination.

Florkin, M. and E. H. Stotz. 1967. Morphogenesis, differentiation and development. Elsevier, New York. A collection of authoritative articles.

Hadorn, E. 1961. Developmental genetics and lethal factors. Methuen, London. A provocative monograph on many of the classical phenomena of developmental genetics, with stimulating discussions of modern approaches.

Loomis, W. F. 1970. Papers on regulation of gene activity during development. Harper and Row, New York. A compilation of key articles related to gene action and cell differentiation.

Ursprung, H. 1967. Development genetics. *In* Annual review of genetics, vol. I. Annual Reviews, Inc., Palo Alto, Calif. A brief survey of modern approaches to problems of cell differentiation.

Weber, R. 1967. The Biochemistry of animal development. Academic Press, New York. A two-volume treatment, by numerous authors, of chemical embryology.

# TWO

## Differential Gene

## Function:

## RNA Synthesis

The products of genetic activity vary greatly in amount in different cell types and in the same cell type at successive stages of differentiation. Thus, genes must be differentially expressed, with respect to both time and place in the organism. The array of gene products in a particular cell type to a large extent characterizes the differentiated state of that cell. A red blood cell, for example, is characterized by its hemoglobin content. The formation of hemoglobin itself requires the translation of messenger RNAs for the polypeptide chains that form the hemoglobin molecule. These messenger RNAs are produced by transcription of the structural loci for the hemoglobin genes. Clearly, the presence of the genetic information for the synthesis of the hemoglobin messenger RNAs and the corresponding globin polypeptides is a prerequisite for the appearance in the erythrocyte of hemoglobin molecules. But it would be an oversimplification to lump the entire sequence of events leading to hemoglobin production under the single term "gene action." A complete understanding of hemoglobin synthesis as an expression of cell differentiation requires an examination of each step in the biosynthetic sequence. In particular, it is necessary to investigate the transcriptional event leading to hemoglobin messenger RNA, the translational event leading to the synthesis of globin polypeptides, and the combination of these polypeptides with heme to form the completed hemoglobin molecule. This last step can be de-

7

scribed as epigenetic because it is not directly specified by the informational content of the DNA. It is at these three levels—transcription, translation, and epigenetic modification—that regulatory control mechanisms operate in determining whether a cell will contain hemoglobin and will therefore have the essential characteristics of an erythrocyte.

A very simple and useful assay system for the study of primary gene function would enable us to measure the synthesis of mRNA and its cognate protein, in vitro, as a DNA dependent reaction. With such an assay system, one could hope to identify the specific control mechanisms responsible for differential transcription and translation. To this date, satisfactory in vitro assay systems of this kind have not been devised, although some progress has been made, which will be discussed later. There are, however, several in vivo assay systems at hand that have already yielded remarkable results.

## The anucleolate mutant of *Xenopus laevis:* genetic lesion and developmental consequences

The African clawed toad, *Xenopus laevis,* is now a common laboratory organism used in many biological and medical investigations. In 1958, investigators at Oxford University discovered a mutant *Xenopus* that contained only one nucleolus in its diploid cells, instead of the usual two. In genetic nomenclature, such animals are designated 1-nu. This nucleolar deficiency does not prevent normal development of the animal. Thus, it was possible to mate two sexually mature individuals of the mutant strain in a cross 1-nu × 1-nu. The resulting embryos consisted of three types of individuals: homozygous wild type (2-nu), heterozygous (1-nu), and homozygous mutant (0-nu), in the Mendelian ratio 1:2:1. The individuals with no nucleoli do not survive beyond an early swimming larval stage.

## The biochemical lesion

When cytochemical and biochemical tests were performed on these lethal embryos, it was observed that they contained less RNA than do normal embryos. Recently, more discriminating biochemical techniques have made it possible to pinpoint the deficiency to ribosomal RNA (rRNA). These results were obtained in an effort to answer the question: "What kind of RNA does a homozygous mutant synthesize as compared to a normal, wild-type, embryo?" Again, two heterozygous (1-nu) individuals were mated. The resulting offspring were allowed to develop to the neurula stage and then were exposed to radioactive $CO_2$. After a 20-hr exposure, the embryos were transferred to a non-radioactive medium and permitted to develop for an additional 2-day period. Then the different genotypes were separated from one another; this is possible with living embryos because the number of nucleoli can be seen readily by phase contrast microscopy. Next, RNA was ex-

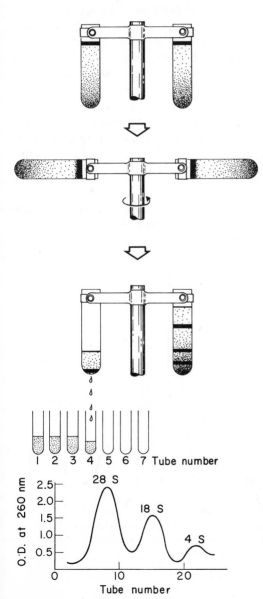

Fig. 2-1 Density gradient centrifugation. Centrifuge tubes are filled with sucrose solutions of different concentrations, so that the bottom of each tube contains a very dense solution that gradually diminishes in density towards the top of the tube. Nucleic acid molecules are then layered on the top of this density gradient and centrifuged. The rate of sedimentation in the centrifugal field depends on the size of the molecules. When the centrifuge is stopped, each size class will occupy a separate invisible zone, or band, in the tube. The bottom of each tube is perforated and the emerging drops are collected in a series of test tubes. These fractions are then analyzed for nucleic acid content by measuring their optical density (O.D.) at a wavelength of 260 nm in a spectrophotometer. The different size classes are designated in Svedberg units (S), after the inventor of the ultracentrifuge. Optical density profile of a common nucleic acid solution after resolution by density gradient centrifugation.

tracted from wild-type and mutant embryos, purified, and separated
into size classes by centrifugation through a density gradient of sucrose.
This technique separates molecules by size (Fig. 2-1). Under these
experimental conditions, three sharply distinguished sizes of RNA are
obtained from a typical vertebrate cell. Identified by their sedimentation constants, they are 4 S, 18 S, and 28 S RNA; the latter two are
components of rRNA. Notice in Fig. 2-2 that all three classes of RNA
are present in both mutant and normal embryos. This result does not
tell us, however, whether these molecules were synthesized by the
embryo or were already present in the egg prior to fertilization. Since
the embryos were exposed to radioactive $CO_2$, which is a precursor of
RNA, it was possible to discriminate between these two alternatives by
measuring the radioactivity of the effluent fractions. Obviously, all
radioactive molecules must have been synthesized by the embryo. As
seen in Fig. 2-2, the normal embryos have synthesized a substantial
amount of RNA of all three size classes, whereas the mutant shows
virtually no synthesis of the 18 S and 28 S RNA, that is, rRNA; notice,
however, that the mutant embryo did synthesize some 4 S RNA.

Fig. 2-2   RNA synthesized in normal and mutant *Xenopus* embryos, optical density profiles
after resolution by density gradient centrifugation. (From D. D. Brown. 1966. J. Exp. Zool.
157:101–114. Fig. 3.)

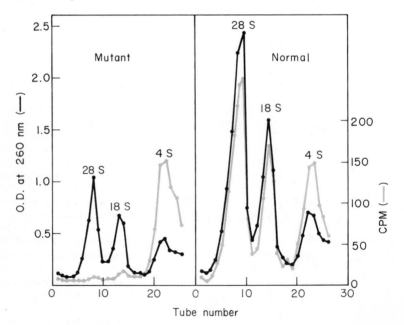

# The genes for ribosomal RNA

These results suggested that in the mutant, the genetic information needed for the production of 28 S and 18 S RNA was absent, or at least dormant, during early embryonic development. Subsequent work proved beyond doubt that the genes for ribosomal RNA are indeed missing in the mutant. Let us examine the type of evidence that was needed to support this statement. In the process of genetic transcription, the linear nucleotide sequence of one strand of DNA is copied in the form of a complementary strand of RNA (Fig. 2-3a). A few years ago, it was discovered that under certain experimental conditions RNA strands could combine with their complementary DNA templates and thus form hybrid molecules consisting of one strand each of DNA and RNA. This phenomenon, known as molecular hybridization or annealing, has found wide application in developmental biology. For the particular case of ribosomal RNA synthesis in wild-type and mutant *Xenopus,* the technique has been used to determine whether the mutant individuals possessed the genes for ribosomal RNA synthesis. Normal double-stranded DNA prepared from wild-type *Xenopus* was dissociated into single strands, bound to a nitrocellulose membrane filter, and incubated with purified radioactive 28 S or 18 S rRNA, obtained from wild-type *Xenopus.* As expected, the RNA molecules combined with their DNA templates to form double-stranded hybrid molecules. These were then separated from the remaining uncombined RNA by passing an eluting fluid through the mixture on the nitrocellulose membrane filter. This filter retains DNA-RNA hybrid molecules, but not free single-stranded RNA. By measuring the radioactivity of the filter after the eluting fluid has passed through it, one can determine the amount of hybridization that has occurred (Fig. 2-3b). When DNA of the anucleolate (0-nu) mutant was used in this experiment, no hybrid molecules were formed, indicating that the annealing sequences complementary to 28 S RNA and to 18 S RNA did not exist in the mutant (Fig. 2-4). Moreover, 18 S and 28 S RNA hybridize to normal DNA independently of one another. The presence of 18 S RNA during an annealing reaction does not interfere with the hybridization of 28 S RNA, and vice versa. This result is to be expected because 28 S and 18 S RNA differ in nucleotide composition, the former having a higher content of guanylic and cytidylic acid (G + C) than the latter.

The chromosomal location of the two genes that code for 28 S and 18 S RNA is known. Soon after the anucleolate mutant was discovered,

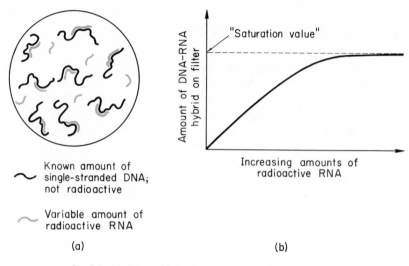

(a)                                        (b)

Fig. 2-3   Nucleic acid hybridization: the saturation experiment.

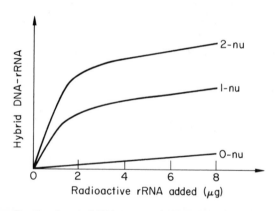

Fig. 2-4   DNA coding for ribosomal RNA in normal (2-nu), heterozygous (1-nu), and mutant *Xenopus* (0-nu).

it was noted that the mutant lacked the secondary constrictions that are normally present on two chromosomes of the normal karyotype (Fig. 2-5). Secondary constrictions on chromosomes in many organisms have been found to be the sites of origin of nucleoli. Generally, the number of secondary constrictions on the chromosomes of the diploid genome corresponds to the number of nucleoli in the cell. Analysis by micromethods of the RNA present in nucleoli reveals that its base composition resembles that found in ribosomal RNA. This observation

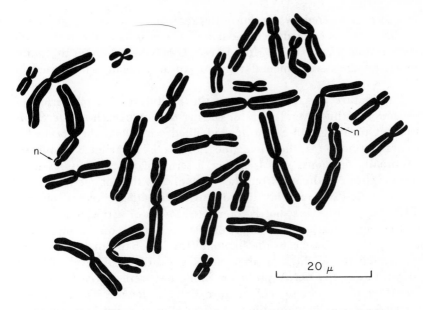

20 μ

Fig. 2-5  Camera lucida drawing of mitotic chromosomes from a normal Axolotl larva. n, nucleolar organizer region. Note that two nucleolar organizers are present in the normal diploid genome. (From H. G. Callan. 1966. J. Cell Sci. 1:85–108. Fig. 1.)

could simply mean that nucleoli accumulate ribosomal RNA. But evidence now shows that the nucleolar organizer is actually at the site of the genes that transcribe ribosomal RNA. Again, by molecular hybridization, it was found that wild-type *Xenopus* has twice as much ribosomal DNA as does the heterozygote (1-nu); the homozygous mutant (0-nu), of course, contains no ribosomal DNA (Fig. 2-4).

## Differential synthesis of rRNA during development

In summary, the DNA encoding 28 S and 18 S RNA is composed of structural genes. Their transcriptional products are stable RNA molecules forming integral parts of ribosomes. Determination of the rates of synthesis of these two kinds of RNA molecules thus provides an opportunity for the study of gene function during the time course of animal development. One could ask the question, for example, whether the two genes were active at all phases of development, or whether, alternatively, there were periods of genetic silence with respect to the production of ribosomal RNA. This question has, in fact, been answered by administering radioactive precursor molecules at different stages of development and by measuring the rate of incorporation of radioactivity into rRNA. It may be helpful at this point to review the developmental stages of an amphibian, schematized in Fig. 2-6.

The mature oocyte of a frog is the product of a long period of oogenesis lasting several months. During this time, each oocyte grows without dividing. Once mature and filled with yolk, the oocyte becomes dormant. At this time, the nucleus is in arrested prophase. Eventually hormonal signals induce a series of transformations leading to release of the egg from the ovary and to the completion of the first meiotic division. This division, together with the second meiotic division, which occurs shortly after fertilization or artificial activation of the egg, leads to a reduction of the 4n chromosome complement of the primary oocyte to the haploid (1n) set of the mature gamete.

To study the synthesis of ribosomal RNA at different developmental stages, radioactive RNA precursor molecules were injected into the dorsal lymph sac of a sexually mature *Xenopus* female. The oocytes in such a female are in many different stages of maturation. If ovulation is induced just a few hours after administration of the label, then the eggs that are ovulated were exposed to the label only during late periods of their development. If ovulation is induced again at some later time, then eggs will be obtained that were confronted with the label during the earlier phases of their maturation.

A different way of achieving the same result is to inject label into the female and then to remove oocytes at different stages of development. Measurements can then be made of the amounts of radioactivity incorporated into the ribosomal RNA synthesized at the various maturation stages. Figure 2-7 summarizes some of the results that have been obtained. Ribosomal RNA is synthesized rapidly by the immature oocyte but apparently not by the ovulated unfertilized eggs. No rRNA synthesis is detectable during the meiotic divisions and very little if any immediately after fertilization. Only at gastrulation is the synthesis of rRNA readily observed again. Although the possibility remains that rRNA synthesis is carried on prior to gastrulation at a rate too slow to have been detected by the means used, most researchers have concluded that the genes for ribosomal RNA are differentially active at various developmental stages.

| | Primary oocytes | | | | | Secondary oocyte |
|---|---|---|---|---|---|---|
| **Developmental stages:** Oogonia | Leptotene Zygotene | Pachytene | Diplotene growing oocyte "Immature oocytes" | "Mature oocytes" | Hormone stimulus → Diakinesis Metaphase I, Anaphase I ("Ovulation") | Prophase → Metaphase (Fertilization) → Anaphase |
| **Time intervals:** hours | very brief | weeks | months | months | hours | hours |
| **Nuclear configuration:** diffuse chromatin | more compact chromatin | condensed chromosomes; "nuclear cap" | Lampbrush chromosomes; nuclear cap disappears | chromosomes condensed | standard | standard |
| **Metabolic activity:** active division | ? | DNA replication; then DNA synthesis in cap | first, DNA-like RNA is synthesized, and then also rRNA | genetic silence | DNA-like RNA synthesized | DNA-like RNA synthesized |

Fig. 2-6  Oogenesis in the clawed toad, Xenopus laevis.

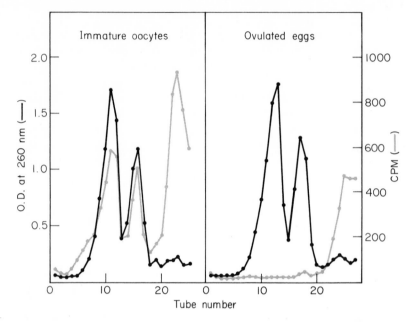

Fig. 2-7  RNA in immature oocytes and mature, ovulated eggs. Note that synthesis occurs in the immature oocyte but apparently stops or proceeds at an undetectably slow pace in the recently ovulated egg. (From D. D. Brown. 1966. J. Exp. Zool. **157**:101–114. Fig. **5**.)

## Control of ribosomal RNA synthesis

If the genes for ribosomal RNA are differentially active at various developmental stages, how is this differential activity brought about? There are at least two quite different mechanisms that could account for the observed facts. First, each gene coding for ribosomal RNA might produce a given number of ribosomal RNA molecules per unit time. From this hypothesis, increased ribosomal RNA synthesis would be a reflection of the number of ribosomal genes coding for RNA, and this number would then have to fluctuate during development. The number would be large in immature oocytes, but virtually zero in the unfertilized egg. Second, each gene for ribosomal RNA could produce either a small amount or a very large amount of ribosomal RNA—that is, the synthetic activity of the respective genes could fluctuate, being very high in the immature oocytes but low in the fertilized egg. In fact, both mechanisms appear to operate during frog development.

## Gene amplification

In the oocyte, at the stage at which ribosomal RNA synthesis is extremely active, the gene for ribosomal RNA exists in a large number of copies. Although the oocyte genome as a whole is only tetraploid, the oocyte contains not 4 nucleoli but of the order of 1000. Each nucleolus not only contains but also actively synthesizes ribosomal RNA. This was shown by autoradiographic analysis of oocytes that had been allowed to incorporate radioactive RNA precursor molecules. Thus, it appears that the genes for ribosomal RNA located in each nucleolus are replicated and released from the nucleolar organizer region of the chromosome during oocyte growth. In a different amphibian, the axolotl, this process has been directly observed through the phase contrast microscope.

Very recently, investigators have succeeded in isolating ribosomal RNA genes from amphibian oocyte nucleoli (Fig. 2-8). Oocyte nuclei

Fig. 2-8  Photograph of ribosomal cistrons transcribing rRNA. Transcription in various stages of completion is evident in the lateral extensions from the central thread of DNA. (From O. L. Miller, Jr. and B. R. Beatty. 1969. Science. 164. Copyright 1969 by the American Association for the Advancement of Science. Courtesy of O. L. Miller, Oak Ridge National Lab.)

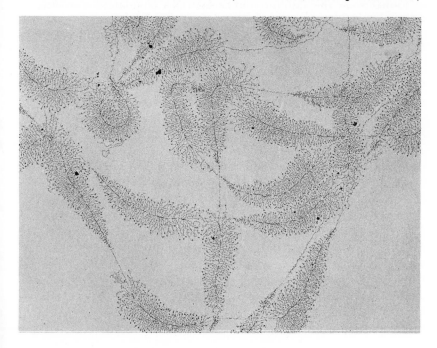

were placed in deionized water and the nucleolar cores centrifuged through sucrose and caught on electron microscope grids. The preparations were then stained with phosphotungstic acid, dried, and photographed with the electron microscope. The figures show featherlike structures, each representing a ribosomal cistron in the process of transcribing some 100 rRNA precursor macromolecules in various stages of completion.

The interpretation of all these observations leads to the conclusion that the oocyte indeed contains many multiple copies of the genes for ribosomal RNA. One should be able to verify this conclusion by molecular hybridization experiments, for a large proportion of the DNA extracted from oocytes should be complementary to ribosomal RNA, larger than that extracted from somatic cell nuclei, each of which contains only two nucleoli. This expectation is fulfilled by the results of several experiments. One must conclude that *Xenopus* oocytes are, indeed, enriched with respect to the genes for ribosomal RNA, by as much as 1000 times compared to somatic cells. This enrichment of genes for ribosomal RNA doubtless accounts for at least part of the extremely rapid rate of synthesis of ribosomal RNA by oocytes.

The technique of molecular hybridization has furthermore enabled us to determine the absolute number of DNA templates for ribosomal RNA per diploid genome. These values were obtained in so-called saturation annealing experiments (Fig. 2-3). In such experiments, the investigator confronts a given amount of DNA with increasing amounts of radioactive ribosomal RNA, and determines the point at which no further RNA-DNA hybrid molecules are formed. This so-called saturation value indicates the proportion of DNA that is complementary to the RNA. In the case of *Xenopus,* about 0.07 micrograms (µg) of 28 S ribosomal RNA are bound for each 100 µg of DNA, indicating that about 0.07% of the genome codes for 28 S ribosomal RNA. A diploid erythrocyte nucleus of *Xenopus* contains 6 picograms (pg) of DNA, or $3.6 \times 10^{12}$ daltons of DNA; 0.07% of this DNA equals $2.5 \times 10^9$ daltons. If this value is divided by the molecular weight of 28 S ribosomal RNA, which is $1.6 \times 10^6$, a value of about 1600 is obtained. Thus, a diploid genome of *Xenopus* must contain roughly 1600 copies of the genes for ribosomal RNA. These multiple copies of genes for ribosomal RNA all seem to be clustered in the nucleolar organizer region of the *Xenopus* genome. We shall consider later whether extreme gene multiplicity is a common characteristic of the genomes of higher organisms.

### Different rates of transcription

Let us return to the question of what accounts for the vast difference in synthetic rates for ribosomal RNA during development. We have noted that the number of DNA templates for rRNA is higher in oocytes than in fertilized eggs, for example. But fluctuations of ribosomal RNA synthesis have been observed during stages in which the number of templates and the number of nucleoli do not vary. Thus the mechanisms of gene control must permit different rates of transcription. If we found that the rate of synthesis of ribosomal RNA per unit of homologous DNA varied, then we could accurately speak of differential transcription. Control at this level may be properly termed "regulation of gene action" and will concern us more in later chapters.

Perhaps the most striking evidence for such differential transcription is the observation that eggs produced by the 1-nu female have the same content of ribosomes as those from wild-type females. Thus, such eggs, with half the number of genes for rRNA, nevertheless produce a normal amount of rRNA.

We have neglected a very obvious, perhaps trivial, explanation for differential RNA synthesis. Could it not be that the embryo simply passess through metabolically active and inactive phases? If so, the observed fluctuation in RNA production would simply reflect the general metabolic state of the cell. This is not the case, however, since other size classes of RNA are not synthesized at the same relative rates at different developmental stages. There are times in development when "heterogeneous RNA" is produced in the absence of detectable synthesis of ribosomal RNA. This heterogeneous RNA will be considered in the following chapter.

## Other types of RNA transcribed during amphibian development

In the previous section, we have restricted our discussion to 28 S and 18 S RNA. These two types make up the bulk of the RNA found in ribosomes. However, a third type, 5 S RNA, is also found in ribosomes. This 5 S RNA is encoded in DNA that is not closely linked to the genes for 28 S and 18 S RNA. But the synthesis of 5 S RNA is nevertheless coordinately controlled with that of 28 S and 18 S RNA. Ribosomal RNA is a very large component of the total RNA in animal cells. It constitutes about 80% of all cellular RNA. On the other hand, it represents only a very small fraction of the total information contained in the DNA of the cell. Surely, the RNA that underlies the differentiated state of the cell is not the ribosomal RNA, but rather that class of RNA often called messenger RNA, or informational RNA, or DNA-like RNA. The base composition of this kind of RNA resembles that of the total cellular DNA but is different from that of ribosomal RNA. In sucrose density gradients, the informational RNA is not distributed in sharply defined classes as is true for ribosomal RNA, but rather this RNA occupies a wide area of the gradient; that is, it is heterogeneous in size.

Through labeling experiments of the kind that were discussed earlier, followed by sucrose density gradient analysis of the newly formed RNA, it was found that heterogeneous or DNA-like RNA is synthesized in the immature oocyte but not in the mature dormant oocyte. At ovulation, synthesis begins again at a low level and becomes abundant in late cleavage. This RNA is synthesized at a rapid rate before ribosomal and 4 S RNA synthesis (Fig. 2-9) becomes readily apparent in the developing embryo.

Characterization of this RNA is difficult because of its size heterogeneity. But here again the anucleolate mutant proved helpful, for this mutant does not synthesize ribosomal RNA. Consequently, all the heavy RNA is DNA-like. In experiments designed to study the nature of the heterogeneous RNA, mutant embryos were exposed to labeled RNA precursor molecules for short periods of time,—for example, immediately following gastrulation. During this time, it was found that heavy RNA, with a sedimentation constant of about 20 S, was syn-

Fig. 2-9  Ribosome-associated, DNA-like RNA synthesized during *Xenopus* development. Synthesis increases rapidly in late cleavage. (From D. D. Brown. 1966. J. Exp. Zool. **157**:101–114. Fig. 12.)

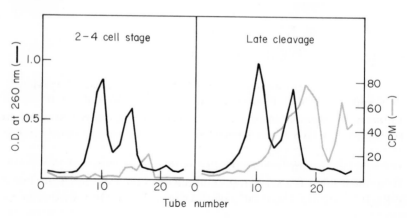

thesized and accumulated until after the primordia of the adult organs formed, that is, up to the tail bud stage of development. By that time, scarcely any radioactive label is present in the embryo in the form of soluble RNA precursor molecules. And yet at the same time, radioactive RNA of smaller size (between 10 S and 20 S) is now observed. This indicates that the large RNA molecules that were originally formed pass through a slow transition in size after the beginning of organogenesis.

In Fig. 2-10 are set forth the major results of the studies of RNA synthesis during amphibian development. Apparently the various size classes and functional classes of RNA are synthesized at different rates during successive stages of development. Such differential RNA synthesis constitutes an impressive demonstration of differential gene function. Amphibians are not the only organisms in which RNA synthesis at various developmental stages has been studied. Similar investigations have been conducted in the nematode *Ascaris,* in insects such as the milkweed bug *Oncopeltus,* in the fruitfly *Drosophila melanogaster,* in sea urchins, in several fishes, and also in mammals. All these studies showed, in essence, that gene function at the nucleic acid level occurs differentially in different developmental stages. But let us not forget that the nature of the individual transcriptional products, except for ribosomal RNA, is obscure as yet. We are certainly not examining the function of individual genes when we measure the total production of heterogeneous RNA, and it is this fraction that contains the molecules that are so crucial for the specification of cellular diversity. Isolation and characterization of individual messenger RNA molecules contained in that broad heterogeneous fraction of RNA is one of the most challenging tasks of present-day biology.

## Tissue specificity of transcription

Thus far, our discussion has centered on the fact that RNA molecules are produced at different rates in different developmental stages. To understand the relevance of such fluctuations for cellular differentiation, we must know where, in the organism, RNA is synthesized, and whether different tissues are different with respect to RNA synthesis. Relatively little is known on this extremely important problem. One experimental design, using the amphibian embryo, involved exposing the dorsal and ventral halves of the embryo to a radioactive precursor of RNA for 1 hr. After that time, the embryos were fixed, embedded in paraffin, and sectioned. The sections were auto-

SYNTHESIS OF RNA SIZE CLASSES DURING DEVELOPMENT

| Stage of development | DNA-like RNA (heterogeneous) | 28 S RNA ($1.6 \times 10^6$) | 18 S RNA ($0.6 \times 10^6$) | 5 S RNA ($3 \times 10^4$) | 4 S RNA ($2.5 \times 10^4$) |
|---|---|---|---|---|---|
| Immature oocyte | + | ++++ | ++++ | ++++ | ++ |
| Mature oocyte | − | − | − | − | − |
| Oocyte during ovulation | + | − | − | − | − |
| Fertilization | + | − | − | − | − |
| Cleavage stages | + | − | − | − | − |
| Late blastula | ++ | + | + | − | ++ |
| Gastrulation | +++ | + | + | + | ++ |
| Post-gastrula stages | ++++ | ++ | ++ | ++ | +++ |

Fig. 2-10 RNA synthesis during Xenopus development. The number of +'s is a rough indication of the relative amount synthesized. Molecular weights are indicated in parentheses.

radiographed and the localization of newly formed RNA was determined. The results indicated that the different germ layers had synthesized different amounts of RNA—that is, there was some tissue-specific RNA synthesis, at least with respect to quantity.

Another approach, potentially much more powerful, involves the use of a modified nucleic acid hybridization technique. In so-called competition annealing experiments (Fig. 2-11), similarities and dissimilarities of primary gene products can, to some degree, be quantitatively determined. As in the ordinary annealing technique described earlier, radioactively labeled RNA is incubated with DNA. But in a competition experiment, this incubation is carried out in the presence of increasing amounts of unlabeled, or competitor, RNA. In principle, if this unlabeled RNA contains annealing sequences very similar to those of the radioactive RNA, then it will compete for annealing sites on the DNA template. In other words, with increasing amounts of competitor RNA added to the incubation mixture, fewer radioactive hybrid molecules will be formed. If, on the other hand, the unlabeled RNA has few or no annealing sequences in common with the labeled RNA, then the presence even of large amounts of cold competitor RNA does not interfere with the number of radioactive hybrids that are formed. The degree of competition is thus some measure of the similarity of RNA molecules.

Fig. 2-11 Nucleic acid hybridization: the competition experiment. See text for procedure.

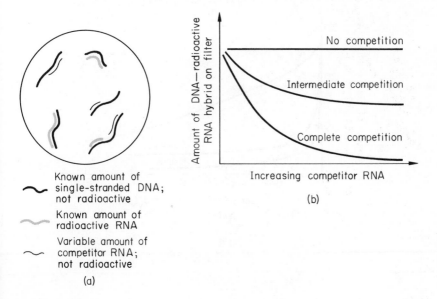

Known amount of single-stranded DNA; not radioactive

Known amount of radioactive RNA

Variable amount of competitor RNA; not radioactive

(a)

In these experiments, radioactive RNA precursor molecules were administered to mice, and the RNA produced during a short time was then extracted from various tissues, such as the spleen, liver, and kidney. RNA was also extracted from animals that had not received injections of radioactive precursors, to serve as unlabeled, competitor RNA. Next, radioactive kidney RNA was hybridized to DNA in the presence of increasing amounts of competitor RNA from various tissues. As is apparent from Fig. 2-12, kidney RNA is the most effective competitor for the annealing reaction between kidney RNA and DNA. Liver RNA is a less effective competitor molecule in this reaction, and so is spleen RNA. This fact strongly indicates that the three tissues do not synthesize identical sets of RNA but that gene function in them is differential. Although the precise nature of the mRNA molecules transcribed in these tissues is unknown, the finding is paralleled by the better-established fact that the enzyme contents of different cell types are also different. Viewed together, these observations at the nucleic acid and protein levels strongly support the concept of differential gene function as the fundamental underlying principle of cellular specialization.

**Fig. 2-12** Diversity of mRNA molecules in normal mouse tissues. Radioactive kidney RNA was hybridized to DNA in the presence of various amounts of non-radioactive RNA from liver, spleen, and kidney. Note that kidney RNA is the most effective competitor. (From B. J. McCarthy and B. H. Hoyer. 1964. Proc. Nat. Acad. Sci. U.S. 52:915–922. Fig. 5.)

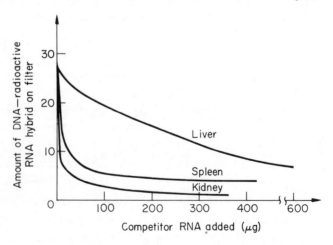

## REFERENCES

Brown, D. D. 1964. RNA synthesis during amphibian development. J. Exp. Zool. **157**:101–113.

Brown, D. D. 1967. The genes for ribosomal RNA and their transcription during amphibian development. *In* A. A. Moscona and A. Monroy [eds.] Current topics in developmental biology, vol. II, p. 48–75. Academic Press, New York.

Brown, D. D. and I. Dawid. 1968. Specific gene amplification in oocytes. Science **160**:272–280.

Callan, H. G. 1966. Chromosomes and nucleoli of the axolotl, *Ambystoma mexicanum*. J. Cell Sci. **1**:85–108.

McCarthy, B. J. and B. H. Hoyer. 1964. Identity of DNA and diversity of mRNA molecules in normal mouse tissues. Proc. Nat. Acad. Sci. U.S. **52**:915–922.

Miller, O. R., Jr. and Barbara R. Beatty. 1969. Visualization of nucleolar genes. Science **164**:955–957.

Wallace, H. and M. L. Birnstiel. 1966. Ribosomal cistrons and the nucleolar organizer. Biochim. Biophys. Acta **114**:296–310.

# THREE

## Differential Gene

## Function:

## Protein Synthesis

In our discussion of RNA synthesis as a measure of gene function, we mentioned an inherent difficulty of investigations carried out at the metabolic level of genetic activity, namely, the problem of recognizing the physiological function of the various kinds of RNA molecules that are synthesized at any particular stage of cell differentiation. Because of this problem, many investigators have chosen to study gene function at the next higher level of expression—that is, at the level of protein synthesis. Much is known about the role of proteins, particularly enzymes, in characterizing the differentiated state of a cell. Although protein synthesis is one step removed from primary gene activity, it is nevertheless a useful indicator of the genetic control of cell differentiation. Different cell types always differ from one another in their protein content, either in possessing different proportions of the same proteins or in containing qualitatively different proteins. Measurements of enzyme activity are commonly equated with measurements of the amount of protein exhibiting that catalytic activity, but this is at best a reasonable inference. Once synthesized, an enzyme molecule may be maintained for varying periods of time in an inactive state. Thus, enzyme activity in a cell may be effectively regulated at the level of gene transcription, RNA translation, and enzyme function.

The nature of the controlling events for the amounts of individual proteins is difficult to determine. Many investigators have therefore chosen to study the rates of total protein synthesis in whole embryos, as a first approximation to the study of gene function. The most comprehensive information of this kind has been gathered from work on sea urchin embryos.

## Protein synthesis during early embryonic development of the sea urchin

In these experiments, sea urchin eggs and embryos at different developmental stages are incubated in seawater containing radioactively labeled amino acids. The labeled amino acids may either be made available throughout the phase of development under study, or, if determination of the rate of protein synthesis is attempted, given as pulses of short duration. Figure 3-1 shows the result of such an experiment performed on embryos of the sea urchin *Arbacia punctulata*. One-hour pulses of radioactive amino acids were given at the times indicated on the abscissa. Incorporation was then stopped by low temperature, the embryos homogenized, and large molecules separated from unincorporated precursor molecules by gel filtration and acid precipitation, techniques that permit the preparative separation of molecules according to size. In the macromolecular fraction, both radioactivity and amount of protein were then determined, and the specific radioactivity of the precipitated proteins plotted as a function of the developmental stages at which the 1-hr pulse had been given (Fig. 3-1). Clearly, the rate of protein synthesis observed in these experiments fluctuates during the course of development. It is close to zero at (and before) fertilization, increases several hundredfold within the first 3 or 4 hr after fertilization, and then declines somewhat before increasing again at the time of gastrulation of the embryo.

## Fluctuation in synthesis: reality or artifact?

When these observations on protein synthesis in sea urchins were first made, three objections were raised immediately. First, it was argued, the rate of uptake of labeled amino acids might fluctuate and be responsible for the observed fluctuations of protein synthesis. Second, the size and composition of the amino acid pool might change

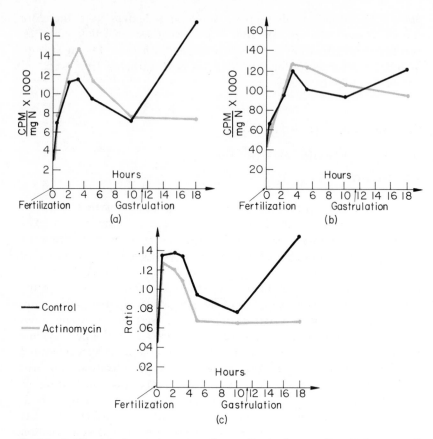

Fig. 3-1    Protein synthesis at various stages of sea urchin development. (From H. Ursprung and K. D. Smith. 1965. Brookhaven Symp. Biol. 18:1–13. Fig. 1.)

with the stage of development. A single amino acid—methionine, for example—might be incorporated into protein at variable rates simply because the pool of methionine fluctuated during the interval studied. Third, the developing embryo might express different levels of metabolic activity that would secondarily affect the rate of protein synthesis.

All these objections have been largely ruled out, although it is true that the rate of uptake of amino acids fluctuates over a wide range during early embryonic development (Fig. 3-1b). This fluctuation alone cannot account for the variation in amino acid incorporation into protein. When a correction is made for the differential uptake, the resulting curve still reveals fluctuations in synthesis (Fig. 3-1c). Yet another way of showing that the rate of amino acid uptake does not determine the rate of protein synthesis involves "preloading" the amino acid pool with radioactive amino acids. In these experiments, amino acids were injected into the body cavity of sea urchins. In this way, the amino acid pool of eggs was labeled long before fertilization. The amount of labeled protein formed in such unfertilized eggs was negligible, although considerable amounts of radioactive amino acids were taken up by the eggs. When such eggs are fertilized, the labeled amino acids contained within them are rapidly incorporated into protein.

To rule out the second objection, it was necessary to determine the pool size of free amino acids. This was done simply by extracting free amino acids at different developmental stages and measuring their relative amounts in an automatic amino acid analyzer. The results of this experiment are shown in Fig. 3-2. Notice that the abundance of certain amino acids in the pool changes in the time period between 4 and 15 hr. But methionine, which was the labeled amino acid used in the experiments shown in Fig. 3-1, does not change appreciably. Thus, change in amino acid pool size does not account for the observed change of incorporation into proteins.

Fig. 3-2 The composition of the amino acid pool during sea urchin development. Values expressed in μmoles/million embryos. (Data from D. J. Silver and D. G. Comb. 1966. Exp. Cell Res. 43:699–700. Table 1.)

|  | Cleavage (4 hr) | Blastula (9 hr) | Mesenchyme blastula (13 hr) |
|---|---|---|---|
| Lys | 1.87 | 1.96 | 1.76 |
| His | Trace | Trace | Trace |
| Arg | 1.40 | 1.56 | 1.61 |
| Asp | 0.21 | 0.22 | 0.19 |
| Thr | 0.67 | 0.93 | 0.75 |
| Ser | 3.26 | 4.84 | 4.46 |
| Glu | 0.67 | 1.21 | 1.06 |
| Pro | Trace | Trace | Trace |
| Gly | 132 | 201 | 214 |
| Ala | 15.1 | 21.6 | 17.6 |
| Cys/2 | Trace | Trace | Trace |
| Val | 0.30 | 0.38 | 0.54 |
| Met | 0.34 | 0.32 | 0.33 |
| Ileu | 0.24 | 0.30 | 0.36 |
| Leu | 0.43 | 0.46 | 0.55 |
| Tyr * | Trace | Trace | Trace |
| Phe * | Trace | Trace | Trace |

* Possibly removed by the charcoal treatment.

The third objection is perhaps the most difficult one to overcome. One investigator has chosen to determine just what kind of protein is being synthesized at different developmental stages. If a general variation in metabolic rate were responsible for the fluctuation in protein synthesis, then one might expect that all protein species would be equally affected at all times. However, this is not the case. When soluble proteins are extracted from pulse labeled embryos and subjected to ion exchange chromatography, several elution peaks are observed at any given time of development. Notice in Fig. 3-3 that protein peak "A" is synthesized in larger amounts than the different peak, "D," during the first hour of development. Seven hours later, the pattern is reversed: now, peak D is produced in larger amounts than peak A. This shift in relative amounts of synthesis of two different classes of protein strongly suggests that the general level of metabolism is not responsible for the fluctuation in the rates of total protein synthesis.

Thus it appears clear that the rates of synthesis do actually vary according to developmental stage. Furthermore, not only are different amounts of protein produced at different stages, but the kinds of proteins produced change. This is particularly noticeable in Fig. 3-3, which shows an elution profile obtained when a 1-hr amino acid pulse was given to embryos 18 hr after fertilization. At this time, at least five clearly recognizable peaks of newly synthesized proteins are evident.

Fig. 3-3   Protein species synthesized at various stages of sea urchin development. (From C. H. Ellis. 1966. J. Exp. Zool. **163**:1–22. Figs. 3, 4, 6.)

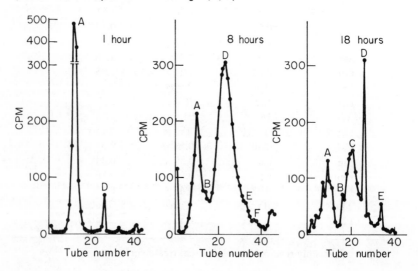

This kind of analysis suffers from one serious reservation, which must be mentioned here and should be kept in mind when interpreting the biological significance of the results. We do not know the metabolic function, nor the precise molecular nature, of those protein species that are eluted from the ion exchange column. Are the various peaks observed different proteins in terms of their primary structure, or do they represent the same protein in different degrees of aggregation, or modified by some conjugated moiety? And just what role in development do these proteins play? Are they enzymes, structural proteins? In short, until we recognize the identity of the molecules whose synthesis we follow, it is impossible even to begin to understand their role in cellular differentiation.

## Controlling elements

Such studies, however, do permit us to examine the mechanisms that control protein synthesis. In the case of the sea urchin embryo, impressive evidence has accumulated that the observed fluctuation in synthetic rate is controlled at the level of both genetic transcription and genetic translation.

### Differential transcription

The general strategy used in these experiments has been to study the effects on protein synthesis resulting from a stoppage of gene transcription. This can be achieved in various ways—for example, by removing the nucleus from eggs by a method developed more than 30 years ago. When unfertilized sea urchin eggs are centrifuged in a medium of a density about equal to that of the eggs, they elongate (Fig. 3-4) and eventually break apart into nucleated and nonnucleated fragments. By a slight modification of the same technique, the two kinds of egg fragments can be harvested in large amounts. Development of such egg fragments can be induced chemically by butyric acid, which also activates intact eggs to develop parthenogenetically. This activation is first manifested in the elevation of a fertilization membrane just as if fertilization had occurred, and then in cleavage of the egg or egg fragment. Obviously, two kinds of embryos develop in such an experiment: one with just the maternal nucleus, and one without any nucleus at all. The latter egg fragments are particularly interesting in the present experiments. They can develop into blastula-like "embryos" consisting of cells lacking nuclei. When the incorporation of amino acids into protein was studied in such anucleate

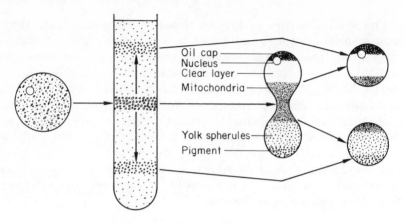

Fig. 3-4    Method of production of nonnucleated fragments of sea urchin eggs.

fragments, it was seen that the rate of synthesis rose sharply after activation, just as in normal development. This finding indicates very strongly that transcription of nuclear information is not required for the initial increase of protein synthesis after fertilization. However, these experiments do not tell us decisively that no transcription is required, because DNA in an egg is not restricted to the nucleus but is also found in the cytoplasm.

Gene transcription can be blocked very effectively by the use of the antibiotic actinomycin. This molecule has been shown to form complexes with guanine in DNA, effectively inhibiting the transcription of genetic information into RNA, without, however, preventing the replication of DNA. As one might expect, treatment of higher organisms with this antibiotic produces drastic metabolic effects. For our present discussion, we note that the rate of increase of protein synthesis after fertilization is not affected by actinomycin (Fig. 3-1). The pregastrula rate increase, on the other hand, does not occur in the presence of actinomycin. There are several ways in which these results can be interpreted. The most obvious conclusion is that RNA synthesis is not required for postfertilization protein synthesis but is required for pregastrula synthesis. From all the evidence available, this conclusion seems most probable. One might also conclude that RNA synthesis does not occur until about the time of gastrulation in normal development. But this conclusion cannot be drawn from the results of the actinomycin experiments alone, and, in fact, has been refuted by experimental data, which show that RNA synthesis does occur as early as the four-cell stage.

Some of this early RNA is DNA-like in base composition and is commonly assumed to be mRNA. At least this heterogeneous RNA does hybridize with a large enough fraction of the DNA to be consistent with the view that it is mRNA. The nature of the heterogeneity of this RNA has been examined by molecular hybridization experiments. In these experiments, radioactive RNA was extracted from the prism stage of sea urchin development. Unlabeled RNA obtained from various other stages was used as competitor RNA. As Fig. 3-5 shows, cleavage stage embryos, and unfertilized eggs as well, contain some RNA that is also synthesized by prism stage embryos. Thus, the sea urchin genome is clearly not completely silent before gastrulation. Moreover, the relative activity of different parts of the genome changes as development proceeds.

These conclusions have meanwhile been verified at the level of protein synthesis. Although the bulk of postfertilization protein is synthesized even in the presence of actinomycin, this in not true for all protein species. Some of the proteins that are synthesized during early development of control embryos are clearly missing in experimental animals that have developed in the presence of actinomycin

Fig. 3-5  mRNA synthesized at various stages of sea urchin development. (From A. H. Whiteley et al. 1966. Proc. Nat. Acad. Sci. U.S. 55:519–525. Fig. 4.)

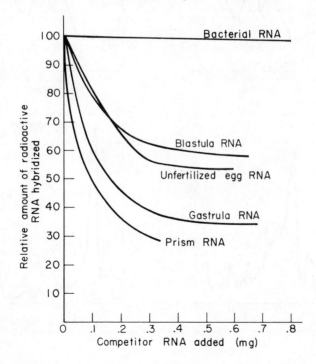

(Fig. 3-6). It is reasonable to assume that the observed low level of RNA synthesis includes mRNA which is responsible for the production of these few proteins. It is also clear that the later, massive synthesis of gastrula proteins is RNA-dependent; actinomycin very effectively prevents the synthesis of these proteins (see Fig. 3-6).

In summary, we may conclude that the great majority of the proteins synthesized around the time of gastrulation in the sea urchin embryo are dependent on concomitant RNA synthesis, but the proteins synthesized immediately after fertilization are not dependent on new RNA synthesis. This latter finding suggests strongly that the "post-fertilization" proteins are synthesized on preexisting RNA templates that were stored in the egg and prevented by some kind of masking

**Fig. 3-6** The effect of actinomycin-D on macromolecular synthesis in sea urchin embryos. See text for description of experiments. (From C. H. Ellis. 1966. J. Exp. Zool. **163**:1–22. Figs. 12, 13.)

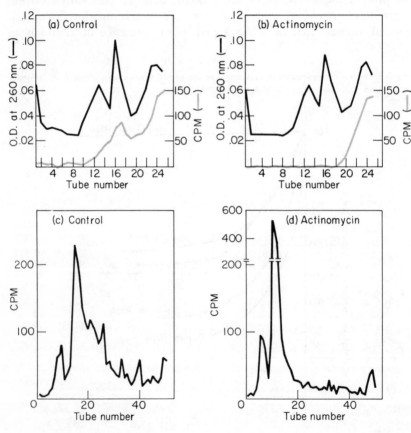

mechanism from participating earlier in protein synthesis. We shall go into more detail on this example of translational control in Ch. 7.

Protein-synthetic patterns have been studied in various other organisms, although perhaps not in as much detail as in the sea urchin. Fluctuations of rates of protein synthesis have been detected during the development of *Drosophila*, for example, and also in early embryonic development of tunicates (sea squirts), and in frogs. There can be no doubt that in a great variety of organisms the rates of gene function undergo substantial fluctuations during embryonic development.

REFERENCES

Ellis, C. H., Jr. 1966. The genetic control of sea-urchin development: a chromatographic study of protein synthesis in the *Arbacia punctulata* embryo. J. Exp. Zool. **164**:1–22.

Gross, P. 1967. The control of protein synthesis in embryonic development and differentiation. *In* A. A. Moscona and A. Monory [eds.] Current topics in developmental biology, vol. II, pp. 1–47, Academic Press, New York.

Hultin, T. and A. Bergstrand. 1960. Incorporation of $C^{14}$ leucine into protein by cell-free systems from sea urchin embryos at different stages of development. Develop. Biol. **2**:61–75.

Monroy, A. 1965. Chemistry and physiology of fertilization. Holt, Rinehart & Winston, New York, 180 pp.

Reich, E. 1964. Binding of actinomycin as a model for the complex-forming capacity of DNA. *In* M. Locke [ed.] The role of chromosomes in development. Academic Press, New York.

Silver, D. J. and D. G. Comb. 1966. Free amino acid pools in the developing sea urchin *Lytechinus variegatus*. Exp. Cell Res. **43**:699–700.

Tyler, A. 1967. Masked mRNA and cytoplasmic DNA in relation to protein synthesis and processes of fertilization and determination in embryonic development. Develop. Biol. **1**:170–226.

Ursprung, Heinrich and K. D. Smith. 1965. Differential gene activity at the biochemical level. Brookhaven Symp. Biol. **18**:1–13.

Whiteley, A. H., B. J. McCarthy, and H. R. Whiteley. 1966. Changing populations of messenger RNA during sea urchin development. Proc. Nat. Acad. Sci. U.S. **55**:519–525.

# FOUR

Differential Gene

Function: Enzyme

Synthesis and

Degradation

In Ch. 3, the differential function of genes in regulating protein synthesis in sea urchin embryos was examined. One apparent deficiency of this work on sea urchins is the inability to identify the individual proteins or to ascertain their roles in the development of the embryo. The conclusions seem sound but are not very satisfying. In the present chapter, we wish to present a detailed examination of the roles of several genes in the synthesis of a specific protein during cellular differentiation.

The most extensively studied protein is hemoglobin. The multiple gene control of this tetrameric protein is now well known, and the early work has been well summarized by Ingram (1963). At least four different genes—$\alpha$, $\beta$, $\gamma$, and $\delta$—encode the subunits of human hemoglobin. Recently, a fifth gene, the $\varepsilon$ gene, has been identified as a contributor to hemoglobin synthesis in the early embryo. These genes function at specific stages in the life history, and their relative activity determines the relative abundance of the different tetrameric forms of hemoglobin. The hemoglobin $\alpha_2\gamma_2$ is characteristic of fetal life but is largely replaced in the adult by $\alpha_2\beta_2$ and to a small extent by $\alpha_2\delta_2$ molecules. These different hemoglobin molecules have somewhat different characteristics that presumably enable them to function efficiently in specific tissue environments, whether of the embryo or of the adult. Considerable information on the structure and function of the hemoglobin genes is now available and has been widely reviewed.

## Lactate dehydrogenase

Another promising protein for the examination of differential gene function during cell differentiation is the enzyme lactate dehydrogenase (LDH). This enzyme has been studied for several years in many laboratories and now provides a rich store of information concerning the genetic and developmental control of enzyme synthesis. LDH is found in all of the cells of vertebrate organisms and in many invertebrates as well. It is apparently indispensable to the life of the organism. Its role in metabolism is to preside over the catalytic interconversion of pyruvate and lactate. Pyruvic acid is a strategically placed intermediate product in the metabolic oxidation of glucose to carbon dioxide and water with a release of energy that can be used in various metabolic activities. The position occupied by lactic acid and by pyruvic acid in the metabolic sequence of glucose breakdown is shown in Fig. 4-1. Note that the metabolism of glucose leads to the release of energy at several steps preceding the formation of pyruvic acid. Energy is also released at later steps, but these steps require oxygen. In the absence of oxygen, the metabolic breakdown of glucose would soon stop if it were not for the presence of the enzyme lactate dehydrogenase. This enzyme, as seen from Fig. 4-1, will bring about the conversion of pyruvic acid to lactic acid with the simultaneous oxidation of the reduced co-factor NADH. The oxidized form of this co-factor, NAD, is essential for the completion of an early step in glycolysis. In the absence of oxygen, the accumulation of lactic acid permits the continued oxidation of NADH to NAD, and glycolysis can then proceed with a liberation of energy. Thus, in the absence of oxygen, LDH serves mainly to generate a temporary storage reservoir of hydrogen (as lactic acid). When oxygen becomes available again, the lactic acid is reoxidized to pyruvic acid and then further oxidized to carbon dioxide and water. Clearly, this enzyme makes it possible for cells to continue to function effectively during transient deficiencies in the availability of oxygen.

One might anticipate that an enzyme so fundamental to cell metabolism as LDH would be little changed during the course of evolution and would have the same genetic basis in all organisms. However, an experimental analysis of this assumption revealed that LDH occurs

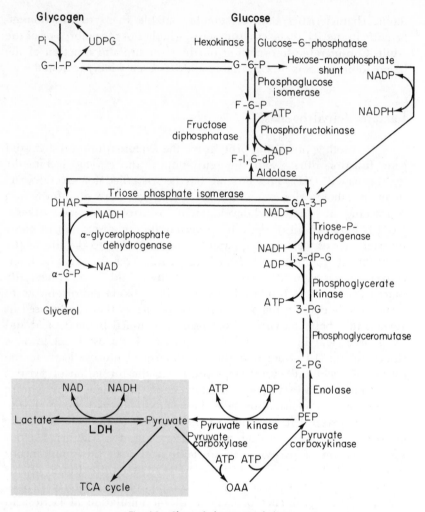

Fig. 4-1    Chart of glucose metabolism.

in multiple molecular forms, as isozymes, in different species and indeed, in the different tissues of a single organism. It is a polymeric enzyme, a tetramer, composed of subunits of essentially the same size. Each subunit has a molecular weight of about 35,000; hence, the complete enzyme has a molecular weight of about 140,000, a rather common size for many polymeric proteins. This information was obtained from the following types of experiments. First, crude homogenates of, for example, mouse skeletal muscle were subjected to electrophoresis in an inert medium, such as starch or acrylamide gels. This technique (Fig. 4-2) is known to separate protein molecules in accord with their

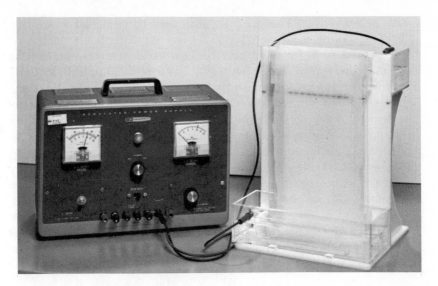

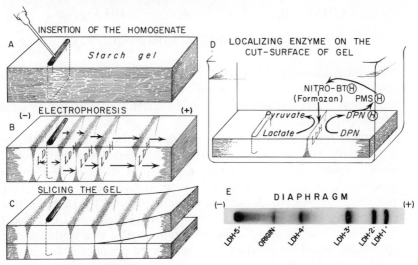

Fig. 4-2 Photograph (top) showing electrophoretic set-up for electrophoresis. A power supply is used to impose a voltage gradient across a starch slab that is held in a vertical position by means of plastic forms. (Bottom) Diagrammed procedure for electrophoretic resolution of homogenates and for staining the gels to reveal the location of LDH isozymes. (Courtesy of W. H. Zinkham.)

net electrical charge. After a mixture of protein molecules has been exposed for some time to the voltage gradient of electrophoresis, each molecular species will be localized, invisibly as a zone, or band, across the gel at some distance from the point of sample insertion in accord with its net charge. The gel block may then be immersed in a histo-chemical staining solution to reveal the location of particular molecular species. The stain for LDH contains lactic acid (the substrate for the enzyme), the required cofactor NAD, and a tetrazolium salt. If LDH is present in the gel, the substrate lactic acid will be oxidized and the tetrazolium salt reduced to a highly colored formazan that precipitates at the site of enzyme activity (Fig. 4-2). Notice in this figure that five colored zones appear after electrophoretic resolution of a homogenate of mouse skeletal muscle. Each of these colored zones identifies an isozyme of LDH. Five isozymes of LDH have been found in virtually all tissues of nearly all mammals and birds so far examined. These five isozymes could be produced by tetrameric combinations of two sub-units, **A** and **B,** formed under the aegis of two corresponding genes. These two kinds of subunits associating in all possible combinations of four would generate the five identified isozymes of LDH. The formulas of these can be written as follows:

$$\text{LDH-1} = B_4, \qquad \text{LDH-2} = A_1B_3, \qquad \text{LDH-3} = A_2B_2,$$

$$\text{LDH-4} = A_3B_1, \qquad \text{LDH-5} = A_4.$$

The designations LDH-1, LDH-2, and so on, are convenient identifica-tions of the five principal isozymes observed after electrophoretic separation. LDH-1 is nearest the anode, LDH-2 is next, and so on to LDH-5 (Fig. 4-2).

To test the tetrameric model for the five isozymes of LDH, we purified the enzyme and separated the isozymes from one another. This is readily achieved through a variety of standard biochemical procedures, including salting-out, column chromatography, and prep-arative electrophoresis. A procedure commonly used for LDH purifica-tion is diagrammed in Fig. 4-3. This procedure yields an LDH that is "electrophoretically pure." A protein stain applied to a gel block after electrophoresis of the purified LDH reveals the presence of five protein bands, corresponding in position to the five forms of LDH; no con-taminating protein bands are present. If this pure preparation is centrifuged through a sucrose density gradient in an analytical ultra-centrifuge, the position of the peak of protein absorption permits a calculation of the molecular weight of the enzyme. It is approximately 140,000. Only one peak is found, indicating that all five isozymes have the same molecular weight. If, in a parallel experiment, the pure preparation is subjected to treatment with agents known to disrupt the three-dimensional structure of functional enzyme molecules, then both the electrophoretic and centrifugal properties of the resulting material are different from those of the native enzyme. The molecular

| Purification step | Total enzyme activity units $\times 10^3$ | Specific activity units/mg | Total yield | Purity % |
|---|---|---|---|---|
| Tissue extraction with buffer | 180.7 | 21.6 | 100% | 2% |
| Ammonium sulfate precipitation | 129.4 | 28.4 | 72% | 3% |
| DEAE cellulose chromatography | 53.6 | 144 | 49% | 15% |
| CM cellulose chromatography | 43.5 | 572 | 40% | 50% |
| Superfine CM cellulose chromatography | 30.0 | 1021 | 28% | 80% |
| Sephadex gel filtration | 20.4 | 1193 | 19% | 100% |

Fig. 4-3   Chart of LDH purification procedure.

weight is reduced to about 35,000 and only two, not five, electrophoretic bands are present when a gel is stained for protein (Fig. 4-4). These two bands are the denatured **A** and **B** subunits of LDH.

Moreover, it is now possible to dissociate the LDH tetramer into its

Fig. 4-4   Zymogram showing banding pattern of LDH isozymes after dissociation in urea. Diagram of banding pattern showing on the left 5 isozymes of LDH separated by starch gel electrophoresis. Each band can be identified by either an enzyme stain or a protein stain. On the right is shown the result after urea treatment of a mixture of all 5 isozymes. After urea treatment the preparation was resolved by electrophoresis and stained for protein. Only 2 bands appeared. No bands appeared after enzyme staining.

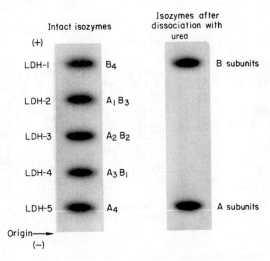

constituent subunits without serious disruption of the three-dimensional structure of each subunit. This is achieved by freezing and thawing LDH molecules in a suitable salt solution. The subunits separate and then recombine at random. Thus, mixtures of the $A_4$ and $B_4$ tetramers can be made to generate all five isozymes in binomial distributions (Fig. 4-5). Presumably, a similar random combination of LDH subunits occurs within the cell, and the final isozyme pattern would then simply reflect the relative abundance of the two kinds of subunits. If these subunits are present in equal amounts, the ratio among the five isozymes will be 1:4:6:4:1. When the relative abundance of the five principal isozymes of LDH in tissue extracts is measured, it is commonly found that the proportions are indeed binomials, but frequently skewed from a 1:4:6:4:1 ratio (Fig. 4-6). Thus, the two subunits are probably synthesized in different amounts. The ratio of isozymes is highly characteristic for each type of tissue. The terminal pattern found in adult tissues develops through a long sequence of isozymic changes from the pattern first seen in the egg. The constancy of these patterns suggests that each of the different isozymes must have its own particular role to play both in adult cells and in the various stages in the differentiation of embryonic cells.

The eggs of certain vertebrate organisms, such as mice, are initially

Fig. 4-5 Dissociation and recombination. On the left is a photograph of a zymogram showing in the first channel LDH-5, in the second channel a mixture of LDH-1 and LDH-5 after dissociation and recombination, and in the third channel, LDH-1. To the right a series of steps are diagrammed illustrating the dissociation and recombination of LDH-1 ($B_4$) and LDH-5 ($A_4$) to yield a binomial distribution of all five isozymes shown in the extreme right. (Courtesy of W. H. Zinkham.)

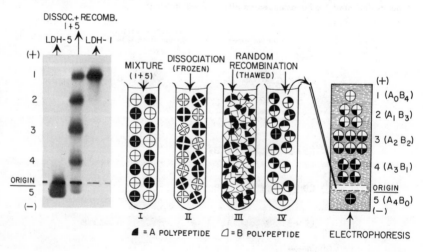

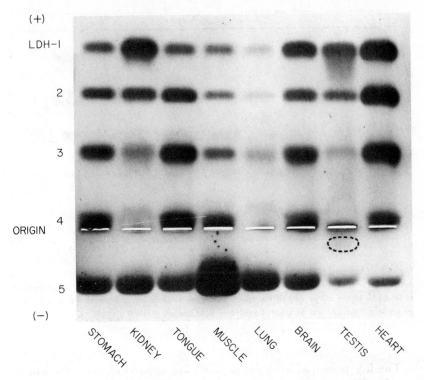

Fig. 4-6  Zymogram showing LDH isozyme banding patterns in eight tissues of the rat.

equipped with LDH-1, the **B** tetramer. Thus, the **B** gene is very active during oogenesis, the **A** gene much less so. As development proceeds, the **B** gene is essentially suppressed and the **A** gene activated. By the time the mouse embryo has reached 9 days of development, the **A** gene is far more active than the **B** gene. Thus, the isozyme pattern is skewed toward the LDH-5 end of the spectrum. With further embryonic development, and particularly at the time of birth and soon thereafter, the **B** gene is progressively activated in many tissues of the body, so that the isozyme pattern shifts again toward the LDH-1 end of the spectrum. These changing patterns reflect the shifting relative activities of the **A** and **B** genes (Fig. 4-7).

It is obvious from an examination of the LDH patterns of various tissues that the relative activties of the **A** and **B** genes vary over an enormous range, at least a hundredfold. Clearly genes are not only inhibited or activated, but once activated, the mechanisms regulating gene expression must control the degree of activity in an exceedingly refined fashion.

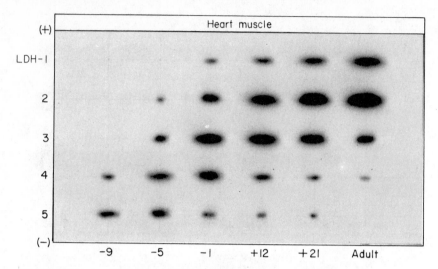

Fig. 4-7   Changing LDH isozyme pattern in heart tissues in the embryonic, juvenile, and adult mouse. Note that the isozyme pattern shifts from predominantly LDH-5 to a predominance of isozymes at the LDH-1 end of the spectrum. In the extreme left channel is the pattern from an embryo 9 days before birth, at the extreme right, the adult heart pattern.

The five principal isozymes of LDH are apparently not sufficient to provide for the requirements of all of the different specialized cells of the organism. At least one kind of differentiated cell in birds and mammals, including human beings, appears to require a different kind of LDH to carry out its metabolic activities. This cell is the spermatocyte. The spermatogonia, progenitors of the spermatocytes, are equipped with the usual LDH composed of **A** and **B** subunits. However, at the inception of spermatocyte differentiation, the **A** and **B** genes appear to be suppressed—that is, turned off—and a third gene, the **C** gene, synthesizing still a different form of LDH, is apparently turned on. The **C** tetramer that results is an LDH molecule with properties that sharply distinguish it from all of the five isozymes composed of **A** or **B** subunits. The **C** gene provides one of the most remarkable examples of the time and cell specificity of gene function. It is quiescent in all the cells of the body except primary spermatocytes. In these cells, it is turned on for only a brief period of time, a few hours, and then turned off again. The evidence for this conclusion comes from genetic studies on pigeons, some of which are heterozygous at the **C** locus. An individual pigeon heterozyous at the **C** locus produces five electrophoretically different isozymes composed of **C** subunits, mutant and normal (Fig. 4-8). The relative abundance of five isozymes in testicular homogenates approximates a binomial distribution of 1:4:6:4:1, thus demonstrating that both alleles were functioning simultaneously in the same cell and to approximately the same extent. After the primary spermatocyte has divided to form the secondary spermatocytes, the **C** genes must be turned off; otherwise, as a result of the reduction division that occurs at this time, the relative abundance of the five isozymes composed of **C** subunits would not reflect a binomial ratio in the homogenates, but would be distorted in favor of larger amounts of the homopolymers. We do not know the reason for the advantage to the spermatocyte of synthesizing this special form of LDH, although the $C_4$ isozymes do have somewhat different kinetic properties and also have broader substrate specificities. This **C** gene has been found in most mammals and birds but not commonly in cold-blooded vertebrates. Thus, the warm-blooded vertebrates possess at least three separate genes for LDH synthesis. Fish also generally possess three or more genes encoding LDH subunits. A few species of fish appear to synthesize a distinct group of isozymes in their gonads, and these isozymes may contain subunits homologous to the **C** subunits of mammals and birds. But this question has not yet been adequately investigated.

However, many fish do possess an additional isozyme of LDH that is

**Fig. 4-8** Diagram showing the electrophoretic resolution of the unique X bands found in the testes of pigeons. Three genotypes are known, designated cc, cc', c'c'. Their corresponding LDH phenotypes are diagrammed with a c'c phenotype shown in the center channel.

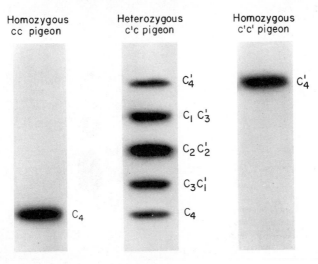

| Homozygous<br>cc pigeon | Heterozygous<br>c'c pigeon | Homozygous<br>c'c' pigeon |
|---|---|---|
| | $C_4'$ | $C_4'$ |
| | $C_1 C_3'$ | |
| | $C_2 C_2'$ | |
| | $C_3 C_1'$ | |
| $C_4$ | $C_4$ | |

found in the cells of the retina and in certain nerve cells of the brain. The LDH polypeptide encoded by this gene, designated the **E** gene, is restricted to adult cells and is not produced in the antecedent embryonic cells. Like the **C** gene in the spermatocytes of mammals and birds, the **E** gene is turned on only at specific stages in development and only in a few kinds of cells. The polypeptides encoded in **E** genes form $E_4$ tetramers and also co-polymerize with **A** and **B** subunits to form additional groups of isozymes. Moreover, in certain species of fish, in which **A** and **B** subunits do not co-polymerize and which therefore form no **A-B** heteropolymers, the presence of the **E** subunit makes possible the formation of tetramers containing **A** and **B** subunits as well as the **E** subunit. Thus, the variety of isozymic forms that can be synthesized in such fish is vastly enhanced through the activation of the **E** gene (Fig. 4-9). Whether this extensive repertory of LDH

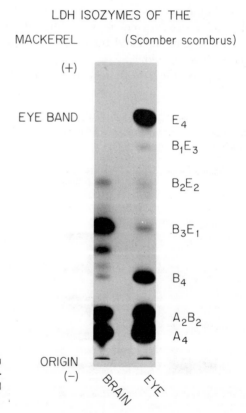

LDH ISOZYMES OF THE

MACKEREL      (Scomber scombrus)

Fig. 4-9 Zymogram showing LDH isozymes containing E subunits in the mackerel eye and brain tissues.

isozymes is essential to the normal physiology of the cells of this fish has not been demonstrated. The widespread occurrence of **E** isozymes in many different species of fish does suggest a significant role for these unusual isozymes in the differentiation and function of specialized nerve cells.

The selective activation of the various LDH genes is ultimately referable to the chemical composition of the cytoplasm, and this composition is, in turn, affected by genes that have regulatory control over the synthesis of one or another of the LDH subunits. One case reported is that of a mutant gene regulating the synthesis of the **B** subunits in mouse erythrocytes. The LDH **B** gene is normally active in nearly all mouse cells, but in mice carrying the mutant gene the activity of the **B** gene is suppressed in erythrocytes and not in other cells (Fig. 4-10). The suppression of the **B** gene in this particular cell type can be shown by the usual techniques of Mendelian genetics to be referable to a single gene. We will discuss the question of regulator genes in higher organisms elsewhere in this book (Chap. 5). It is especially important to note that the activity of this regulatory gene is re-

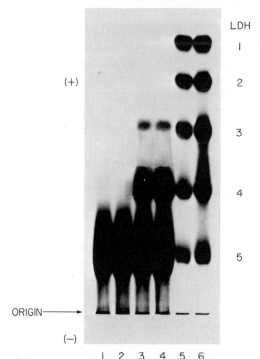

Fig. 4-10 Zymogram showing genetic regulation of LDH-B subunit synthesis in the mouse. (Courtesy of T. B. Shows and F. H. Ruddle.)

stricted to a single cell type, namely the erythrocyte, and thus its action is uniquely dependent upon a highly specialized cellular phenotype. Thus, regulatory genes, like structural genes, exhibit cellular specificity.

In long-lived vertebrate cells, in contrast to bacterial cells, the repertory of protein molecules in the cell at any one stage is a function not only of synthesis but also of degradation. Thus, with reference to the isozymes of LDH, the repertory in the cell at any one time represents a balance between the processes of synthesis and degradation. As the pattern of isozymes changes during cellular differentiation, the obsolescent isozymes must be differentially removed. Evidence to demonstrate this differential degradation has recently been presented. By isotopic labeling experiments in the rat, it was shown that the rate of degradation of LDH-5 is both different and characteristic for each cell type (Fig. 4-11). The rate of synthesis is also characteristic for each cell and dependent upon the rate of gene activity. Synthesis is analogous to a zero-order reaction. On the other hand, the degradation of LDH molecules is a first-order reaction. The steady state level of LDH in any cell can be increased or decreased by altering either the rate of synthesis or the rate of degradation. Each process can be varied over an enormous range and is dependent upon the state of differentiation of the cell.

From this brief account of the enzyme lactate dehydrogenase, it is clear that not only must genes for a given protein be turned on before that protein will be synthesized in a cell, but once activated, the magnitude of gene function is regulated in a very fine-grained fashion over a wide range. Moreover, the product of the gene is itself subject to degradation at many different rates. Whether the primary regulation of gene activity is an all-or-none phenomenon with subsequent modulation of gene activity residing in other mechanisms, or whether

Fig. 4-11   Metabolism of LDH-5 in rat tissues.

| | Synthesis p moles/day/g | Degradation fraction of LDH/day | Quantities µg/g tissue | Half-life (days) |
|---|---|---|---|---|
| Heart muscle | 15.4 | 0.399 | 5.4 | 1.6 |
| Skeletal muscle | 31.0 | 0.018 | 240.0 | 31.0 |
| Liver | 65.0 | 0.041 | 224.0 | 16.0 |

the mechanism that primarily activates or inhibits the genes also functions to specify the degree of activity, is one of the great unanswered questions of contemporary research in developmental genetics. Lactate dehydrogenase may prove to be especially useful in answering this question, but certain other enzymes are also sensitive indicators of differential gene function.

## Xanthine dehydrogenase in
## chick liver development

Xanthine dehydrogenase (XDH) is one of the many enzymes that have been observed to increase suddenly at the time of hatching in the chick (Fig. 4-12). In keeping with current views on gene function, one would conclude that this is a consequence of the underlying gene being "turned on" at the time of hatching. Of course, this assumption is not a priori valid. The same effect—that is, sudden increase in specific enzyme activity—could result from an onset of

Fig. 4-12 Developmental curve of XDH in chick liver. (From G. Murison. 1969. Develop. Biol. 20:518–543. Fig. 1.)

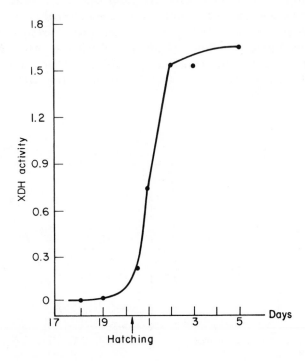

translation of preexisting messenger, from a differential accumulation of enzymes produced elsewhere, from activation of preexisting enzyme, or from a cessation of enzyme degradation. In the present case, activation has been ruled out as the underlying mechanism. If a chicken liver extract containing high levels of activity is mixed with one obtained from an earlier stage containing no activity, the activities of the mixture are strictly additive. This is also true in the reciprocal experiment. Consequently, the high values observed at later stages are not due to activation of dormant enzyme nor are the low levels of earlier stages a consequence of inhibition by inhibitors.

The question still remains, however, whether the increased XDH activity is due to increased production of enzyme protein. To answer this question, we purified XDH, injected the resultant protein into a rabbit, and obtained specific anti-XDH antibodies. The potency of these antibodies was tested in an immunotitration experiment. Antibody was added to increasing volumes of purified XDH. The resulting precipitates were centrifuged, and XDH activity in the supernatant measured. Figure 4-13 shows that up to 50 units of XDH activity were precipitated in the titration experiment by 30 microliters (μ liter) of antibody. Beyond this so-called "equivalence point," XDH began to appear in the supernatant. Thus, in this particular experiment, each microliter of antibody can precipitate about 1.6 units of XDH. With this knowledge, it is possible to investigate whether stage-specific differences in enzyme activity are correlated with correspondingly increased amounts of immunologically active material. All one does is repeat the same titration experiment, using XDH from different developmental stages. For each stage, the enzyme was partially purified

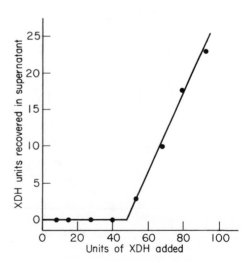

Fig. 4-13   Immunotitration of purified
XDH. (From Murison. Fig. 6.)

and the equivalence point determined. Notice in Fig. 4-14 that for the various XDH preparations used the equivalence points are always the same. There is an absolute correspondence of the catalytic and immunological reactivity of the enzyme at all stages, indicating that the increase in XDH activity is indeed paralleled by an increase in the numbers of molecules of XDH.

This type of approach may be carried one step further. One may ask whether the increased amount of enzyme is brought about by an increased rate of enzyme synthesis or, alternatively, by enzyme stabilization. We shall discuss the combined radiolabel-immunoprecipitation technique used in such analyses elsewhere in the book, in connection with a hormone-induced enzyme increase (p. 83). Experiments analogous to those described in the hormone system were also performed in the present case. Radioactive amino acids were made available to the chicken at various stages of development, and the radioactivity of material precipitated with specific antibody was then determined. The results obtained in these experiments indicate that the accumulation of XDH during chick liver development does indeed result from an increase in the rate of XDH synthesis. Once a high level of XDH is reached, further increase is prevented by the onset of enzyme degradation. This degradative mechanism, whatever its nature, in this case appears to be decisive in establishing steady-state conditions.

## Aldehyde oxidase in *Drosophila*

Another major question still remains unsolved. It is the problem of cell autonomy in enzyme synthesis. The experiment just reported does not test whether the enzyme is actually synthesized by liver

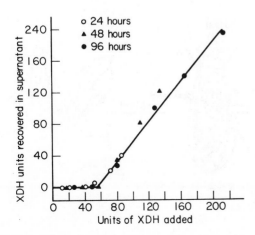

Fig. 4-14 Immunotitration of XDH obtained from three stages of development. (From Murison. Fig. 8.)

cells rather than being accumulated from elsewhere in the organism. In the absence of organ culture experiments, mere observation of enzyme presence (even presence of labeled enzyme) in a tissue does not constitute proof of autonomous synthesis. In fact, several cases are known of proteins manufactured in one tissue accumulating in another. Yolk proteins of insect eggs, for example, are manufactured in "liver" cells and only later transported to the egg. The vertebrate liver itself is known to produce many serum proteins for export. The mere presence of serum proteins in the blood is not, of course, taken as evidence that these proteins are synthesized in the blood.

Several methods, including organ culture, are available to discriminate between autonomous synthesis and importation as sources of any particular protein in a cell. For example, organ primordia may be transplanted between organisms differing in enzyme genotype. The enzyme aldehyde oxidase in *Drosophila* may serve as an example. Several mutants have been found for this enzyme. One of them, "aldehyde oxidase negative," lacks the ability to produce aldehyde oxidase. When a larval ovarial primordium of a wild-type genotype was implanted into a host larva of the aldehyde oxidase negative variety, and the differentiated implant recovered after metamorphosis by dissection from the adult host, the implant was found to contain a normal amount of aldehyde oxidase activity (Fig. 4-15). The implanted ovary must have produced this enzyme by itself, because the host organism was genetically precluded from synthesizing any of this enzyme.

Fig. 4-15   Aldehyde oxidase activity in implanted tissue. The donor ovarian tissue is wild-type; the host is a mutant unable to synthesize aldehyde oxidase. After residing in the host until adulthood the implanted ovarian tissue was removed and tested for the enzyme. The tissue contained the enzyme and must have synthesized it autonomously. (From H. Ursprung et al. *In* Park City Symposium RNA 1969.)

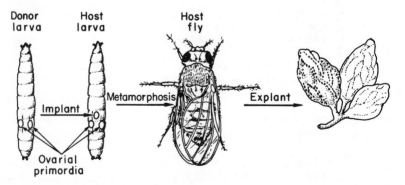

## Hormone-induced enzyme synthesis in plants—the density labeling technique

Several enzymes are known to increase sharply in specific activity during germination in plants. Malate synthetase, for example, has been observed to increase in activity in cotyledons of germinating peanuts. Does this enzyme preexist in some inactive form in the dormant seed, only later to be activated? Or, alternatively, is new enzyme synthesized from a pool of protein precursors? To discriminate between these two possibilities, peanut cotyledons were grown in density-labeled water ($H_2^{18}O$, or $D_2O$). A total protein extract of such cotyledons was then subjected to high-speed centrifugation in a density gradient of cesium chloride. Malate synthetase activity was determined in each fraction recovered from the centrifuge tubes. Figure 4-16 shows that

**Fig. 4-16** Synthesis of malate synthetase in presence of heavy water. Equilibrium distribution in a cesium chloride density gradient of malate synthetase present in a mixture of crude extracts from peanut cotyledons grown in 100 percent $D_2O$ or 100 percent $H_2O$. The $H_2O$ enzyme alone has a buoyant density of 1.270. The greater density of the 100 percent $D_2O$ enzyme implies de novo synthesis from deuterated amino acids. (From C. Longo. 1968. Plant Physiology. 43:660–664. Fig. 6.)

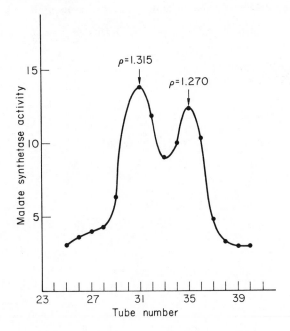

two density peaks with malate synthetase activity were present, one at 1.315 and the other at 1.270, when a mixture of extracts from cotyledons grown in heavy or ordinary water was centrifuged. Obviously, the enzyme activity associated with the heavy fraction is newly synthesized enzyme.

## REFERENCES

Fritz, Paul J., Elliot S. Vesell, E. Lucile White and Kenneth M. Pruitt. 1969. The roles of synthesis and degradation in determining tissue concentrations of lactate dehydrogenase-5. Proc. Nat. Acad. Sci. U.S. 62:558–565.

Ingram, Vernon M. 1963. The hemoglobins in genetics and evolution. Columbia University Press, New York.

Markert, C. L. 1963. Lactate dehydrogenase isozymes: dissociation and recombination of subunits. Science 140:1329–1330.

Markert, C. L. 1968. The molecular basis for isozymes. Ann. N. Y. Acad. Sci. 151:14–40.

Markert, C. L. and H. Ursprung. 1962. The ontogeny of isozyme patterns of lactate dehydrogenase in the mouse. Develop. Biol. 5:363–381.

Markert, C. L. and G. S. Whitt. 1968. Molecular varieties of isozymes. Experientia 24:977–991.

Shaw, Charles. 1969. Isozymes: classification, frequency, and significance. Intern. Rev. Cytol. 25:297–332.

# FIVE

## Regulation of Gene Function

### DNA content of vertebrate cells

How many genes does an organism have? Since genes are clearly important in the differentiation of cells and also determine the characteristics of specialized adult cells, then one might expect to find some simple correlation between the total amount of DNA in the cells of an organism and the variety of differentiated cells making up that organism. Stated somewhat differently, the informational content of an organism should bear some direct relationship to the total amount of information required to make that organism. Analyses of the total DNA contents of various organisms indicate a general conformity between the complexity of the organism and the amount of DNA it contains, just as one might expect. However, certain notable exceptions exist, and these pose serious problems for our understanding of the organization of DNA in chromosomes, and of the role of genes in determining the individual characteristics of cells. Figure 5-1 shows the relative DNA content of the genomes of a variety of organisms ranging over the whole spectrum of evolutionary types, from viruses at one end of the scale to vertebrate organisms at the other. To understand the meaning of these various amounts of DNA, it is essential to understand the coding capacities of DNA. We know that three nucleotides are required to code for a single amino acid, that 64 codons can be formed from the four nucleotides found in cells, and that codes must be available for at least 20 amino acids and per-

| | |
|---|---|
| T2 bacteriophage | 0.00006 |
| *Escherichia coli* | 0.0014 |
| Sponge (*Dysidea crawshagi*) | 0.016 |
| Tunicate (*Ciona intestinalis*) | 0.056 |
| Coelenterate (*Cassiopeia*) | 0.096 |
| Amphioxus (*Amphioxus lanceolatus*) | 0.167 |
| Echinoderm (*Lytechinus pictus*) | 0.26 |
| Flatfish (*Pleuronichthys verticalis*) | 0.23 |
| Trout (*Salmo gairdneri*) | 0.80 |
| Mouse (*Mus musculus*) | 1.00 * |
| Frog (*Rana pipiens*) | 2.17 |
| Newt (*Triturus viridescens*) | 13.00 |
| Salamander (*Necturus maculosus*) | 27.89 |
| Lungfish (*Lepidosiren paradoxa*) | 35.40 |

\* ($1.00 = 7 \times 10^{-12}$g DNAg: the DNA content of a mouse diploid nucleus.)

Fig. 5-1  Relative DNA contents from Feulgen photometry of 14 different species ranging from bacteriophage to vertebrates. All species are diploid except the bacteriophage and E. coli.

haps for some punctuation. When all of these considerations are added together, it is evident that much redundancy is apparently built into the coding system of DNA. It has been shown that almost every amino acid is coded by more than one triplet of nucleotides. Thus, the code is degenerate, or, more explicitly, the same amino acid can be coded by different sequences of nucleotides. If we assume that an average size protein has a molecular weight of about 30,000, then the average number of amino acid residues per protein molecule is about 300, because the average weight of an amino acid is about 100. This number of amino acids can be encoded by 900 nucleotides—in round numbers, 1000 nucleotides per protein. If we look now at the nucleotide or DNA content of various organisms, we find that the virus T2 can code for 200 proteins of average size. The bacterium *Escherichia coli* has enough DNA to code about 4500 proteins, and mammals, including human beings, contain enough DNA to code for 3.2 million.

This range of values fits in a rough way the relative complexity of these organisms. However, this generalization that complexity equals DNA content does not apply within the vertebrates as a distinct group. The range of DNA values among vertebrates varies as much as 150 times. Flatfish, such as the flounder, contain about one-fifth as much DNA as do mammals. Perhaps this value is not surprising, but lungfish contain 150 times as much DNA as flatfish and 35 times as much

DNA as mammals. These values are astonishing. Moreover, various amphibians, such as the salamander *Amphiuma*, contain 25 times as much DNA as does a human being. In terms of coding requirements or informational content, it seems almost inconceivable that more information would be required to code for a lungfish or a salamander than would be required to produce a human being. This is one incongruity in the data that requires explanation.

The second important point to note is that the amount of DNA in mammalian cells is sufficient to code for at least 1000 times as many proteins as have so far been identified. Clearly we have not recognized all the different proteins synthesized in mammals, but it seems highly unlikely that for each identified protein at least 1000 more remain to be discovered. Our knowledge of the biochemistry of organisms is far from complete, but it is complete enough to generate considerable confidence that this enormous amount of DNA cannot be required to code for as yet undiscovered enzyme molecules required in cell metabolism. This is the second incongruity that needs to be explained.

DNA is located primarily in the chromosomes, but minor amounts are also found in cell organelles, such as the mitochondria, chloroplasts of plants, and in the basal bodies of the pellicle of protozoan ciliates. The total amount of DNA in cell organelles other than the chromosomes does not affect significantly the analysis of the relative amounts of DNA of different organisms. Since these must vary among vertebrates over a 150-fold range, an explanation must be sought in terms of chromosome structure or chromosome multiplicity. Several possibilities may be considered. The first is polyploidy. Many organisms do exhibit polyploidy, and the total DNA content of polyploid cells does vary exactly in accord with the degree of ploidy. Haploid cells contain half as much DNA as diploid cells, which, in turn, contain half as much as tetraploid cells. However, a direct examination of the karyotype of the cells of vertebrate organisms fails to support the conclusion that polyploidy could account for the difference in amounts of DNA. In fact, few vertebrates give any evidence for any polyploidy, let alone an amount that would account for a 150-fold difference in the quantity of DNA.

A second possibility is polyteny. Polytenic chromosomes are characteristic of certain dipteran insects, such as the fruit fly, *Drosophila*. The DNA of polytene chromosomes replicates, but the strands of DNA remain attached side-by-side to one another so as to produce a giant cable. More than 1000 strands of DNA may compose a single polytene chromosome. Polyteny is much more difficult to exclude as an explanation for the variable amounts of DNA in vertebrate cells. However, analyses of lampbrush chromosomes in amphibian oocytes clearly indicate that these chromosomes contain only a single DNA double helix. Moreover, the segregation of mutant genes is quite inconsistent with a polytene structure of the chromosomes in gametes.

A third possibility is a linear multiplicity of genes. Many genes might be present in multiple copies attached end-to-end to make up the overall linear structure of the chromosome. Such linear multiplicity might be characteristic to some degree of every gene or might be restricted to only a portion of the genome, presumably the portion that must respond to heavy demands for rapid function sometime in the life of the cell. Thus, in the lungfish each gene might be present in 150 copies in each haploid set of chromosomes, as compared with the flatfish, in which each gene might be present as a single copy. Alternatively, the multiplicity might be different for different genes. Moreover, if differential multiplicity of genes exists at all, it may vary with each tissue of the same organism so as to provide a basis for different quantitative activities of genes during cell differentiation.

Before taking up the specific evidence relating to the possibility of linear multiplicity of genes, we should point out certain implications. First, additional copies of genes would surely permit a more rapid coding for messenger RNA to accelerate the synthesis of certain proteins. However, the existence of multiple genes of a single kind along the chromosome would surely expose the genome to continued divergence by mutation. Each copy of a gene would be subject to independent mutation, and thus, over a period of time, mutations would surely accumulate. Selection pressure to eliminate mutant genes would be greatly reduced as the number of copies increased. The existence of many proteins in isozymic form (Ch. 4) may suggest that just this kind of multiplicity does exist. There is, however, no evidence whatever to suggest that the multiplicity of proteins as seen in isozymes is at all equal to the possibilities presented by the very large amounts of DNA found in vertebrates.

We addressed ourselves to this issue by examining the multiplicity of various kinds of proteins extracted from lungfish tissues as compared with the corresponding multiplicity of proteins in flatfish. The difference in DNA content between these two fish is perhaps the greatest among all vertebrates, being 150-fold. We can briefly summarize all these investigations by pointing out that the proteins of the two fish were not impressively different in multiplicity. If the cells of lungfish contain 150 copies of every gene, or many more copies of fewer genes, as compared with flatfish, then such multiplicity is clearly not expressed in mutant forms of proteins. Although we obviously could not examine all proteins by our electrophoretic methods, our examination was extensive enough to have revealed any multiplicity if it were a general phenomenon. These results drive us to the conclusion that linear multiplicity, which we believe exists, must be organized in such a fashion that only one copy of each kind of gene is transmitted from one generation to the next.

Several investigators have offered suggestions as to how this might be achieved. One of the hypotheses presented assumes a "master" gene controlling the structure of derivative "slave" genes. This postulate requires a mechanism for correcting mistakes or mutations in all the subordinate genes. Thus, in effect, only one gene would carry information between generations. Gene multiplicity based on such an arrangement would not produce a multiplicity of proteins.

This hypothesis is adequate in logic but requires an extraordinary repair mechanism for which we have no evidence. We prefer a different hypothesis based on the known behavior of the ribosomal cistrons. These cistrons are not only linearly reiterated along the chromosome but are also, in oocytes, replicated in many additional copies that are not passed on to the next cell generation. Thus, if these extra copies were mutated, the consequence would be undetectable and transient. We do not understand the chromosomal organization that makes such gene amplification possible, but the phenomenon does exist and may be taken as a model for other genes without invoking novel gene behavior.

In this hypothesis of gene multiplicity (Fig. 5-2), all extra gene copies would be ignored at each cell division, or at least during gametogenesis. Thus, only the primary gene would be replicated, and only mutations appearing in it would become apparent in succeeding generations.

Recently, an examination of gene multiplicity in a wide variety of organisms has been made by means of DNA-annealing techniques. The strands of DNA making up a double helix can be separated from one another by heat or alkali denaturation. They will reassociate again into double-stranded complexes when incubated together under appropriate conditions. To assess gene multiplicity, the DNA is first extracted then denatured and immobilized as single-stranded DNA in some medium such as agar. Next, the immobilized DNA strands are exposed to single-stranded fragments of radioactive DNA obtained from the same organism. Satisfactory reproducible associations of the DNA strands are achieved if the cation concentration, the temperature of incubation, and the concentration of the DNA strands, as well as fragment size of the DNA, are carefully adjusted. The degree of reassociation can be measured in several ways. When one of the strands of DNA is radioactive, it is possible to measure the total reassociation by measuring the amount of radioactivity bound to the immobilized nonradioac-

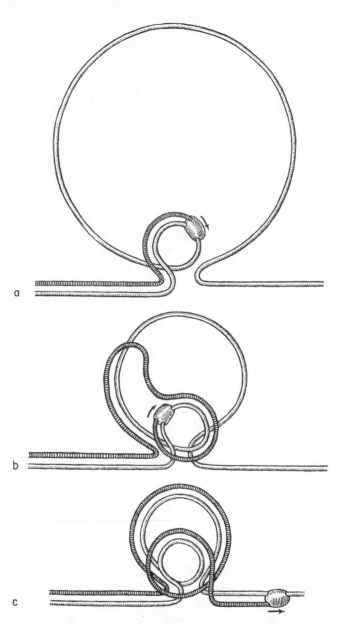

Fig. 5-2 Three diagrams illustrating the hypothesis for generating reiterated sequences of DNA from a single, primary gene. After dissociation of double stranded DNA, each strand is thrown into loops as shown in the diagrams, the small loop representing the primary gene and the large loop representing repeated copies. Because of the arrangement of the loops, the DNA polymerase as it moves (a) along the single strand of DNA is hypothesized to "jump the track" where the strand crosses itself, that is, at the end of the sequence making up the primary gene, and thus it continues to replicate more copies of the primary gene sequence (b) while ignoring the repeated sequences. When enough such copies have been generated to anneal with the original length of repetitive sequences, the DNA polymerase again "jumps the track" and begins (c) to replicate the next gene.

tive strands of DNA. The absorption of ultraviolet light may also be used as a measure of complex formation, because dissociated DNA absorbs more ultraviolet light than the reassociated, annealed, double-stranded DNA. After reassociation has occurred, it is possible to wash away the nonannealed DNA from the immobilized fragments. If the annealing reaction is performed in solution, the annealed duplexes may be separated from single-stranded fragments by passing the mixture through a calcium phosphate gel, to which the reassociated DNA adheres while the single-stranded DNA passes through unabsorbed. During the annealing process, the complementary pairing of bases need not be completely precise. A certain amount of discordance between the sequence of nucleotides in the two strands of associating DNA is permissible. However, the annealing temperature must be lower if base paring is not precise. In fact, there appears to be a general relationship between the effective annealing temperature and the precision of the base paring. The annealing process depends upon molecular collisions between complementary strands of DNA, and the rate at which this occurs depends upon the concentration of the annealing strands. Thus, if all DNA sequences are unique—that is, if each gene is present in the genome just once—then the rate of annealing will be inversely proportional to the size of the genome and will be exceedingly slow for genomes that are the size of those found among vertebrates.

This fact makes it possible to use the annealing techniques to measure gene multiplicity. When reassociation experiments were performed with the DNA obtained from a wide range of organisms, it was found that a reasonably direct proportionality between the rate of annealing and the size of the genome did prevail for lower organisms. But the rate for vertebrates was much too high. A part of vertebrate DNA annealed much more rapidly than would be expected if all sequences of the DNA were unique. These results suggest that certain sequences of DNA are repeated many times in the genome of higher organisms. The results shown in Fig. 5-3 indicate that about 40% of the DNA from calf tissues consists of sequences repeated between 10,000 and 1 million times. In fact, the kinetics of the annealing process with mouse DNA demonstrate that many different degrees of repetition of the DNA must be present, varying from perhaps 100 copies to as many as 100,000 or even 1 million copies of a single nucleotide sequence (Fig. 5-4). From the point of view of cellular differentiation, it is important to know whether the degree of repetitiveness of the DNA varies from one tissue to another or at different stages in cell differentiation. The data are not very sensitive, but there is no evidence for variation in the relative amounts of repeated sequences in

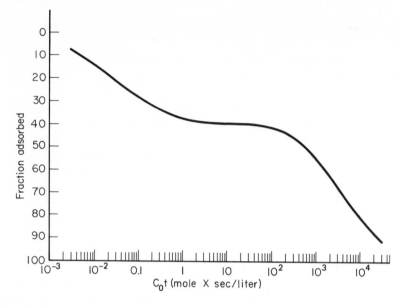

Fig. 5-3 Kinetics of reassociation of calf-thymus DNA. (From R. J. Britten and D. E. Kohne. 1968. Science **161**:529–540. Fig. 3. Copyright 1968 by the American Association for the Advancement of Science.)

Fig. 5-4 Gene multiplicity in the mouse genome. Spectrogram of the frequency of repetition of nucleotide sequences in DNA of the mouse. Relative quantity of DNA plotted against the logarithm of the repetition frequency. These data are derived from measurements of the quantity and rate of reassociation of fractions separated on hydroxyapatite. The dashed segments of the curve represent regions of considerable uncertainty. (From Britten and Kohne. Fig. 12.)

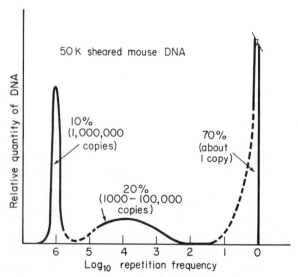

different tissues during development. Although there is no evidence for differential multiplicity of genes in different cells of the same organism (except for oocytes), there is nevertheless the possibility of differentially activating parts of the repeated genes in any cell. It does seem clear that the genomes of higher organisms contain many genes that are represented singly and others that recur from 1000 to perhaps 1 million times per cell. A large part of the total genome is therefore repetitious.

In addition to the DNA of the chromosomes, DNA is also found to a minor degree in other parts of the cell. We now know that DNA is found in each mitochondrion and perhaps in the basal bodies of cilia, but these extra chromosomal sources of DNA do not add very much to the total. In fact, they make up less than 1%, and therefore can scarcely affect the conclusions drawn from the annealing experiments. Nevertheless, it is important to note that the mitochondrial DNA also represents a class of DNA in which the individual genes in a cell must be repeated thousands of times, at least as many times as there are mitochondria. The evidence available strongly suggests that mitochondria are self-reproducing and are passed on from one cell to the next at the time of cell division. Thus, the mitochondrial genes, if subject to mutation, should gradually change so that the mitochondria of an organism would become heterogeneous, unless they were all recent derivatives of the same mitochondrion. Therefore, an examination of the mode of inheritance of mitochondria might give us some insight into the possibility of mitochondrial heterogeneity arising through mutation.

During gametogenesis, the chromosomal genes of an organism are rearranged through independent assortment and by crossing over between homologous chromosomes during meiosis. The genetic contribution of the DNA of mitochondria to the total genetic information of the gamete is still unclear, although some data have been obtained. During fertilization, chromosomal genetic variability is obviously enhanced by the combination of the genomes of egg and sperm. Does a mixing of genetically different mitochondria also occur during fertilization?

Both sperm and egg of higher organisms contain mitochondria, each with its own complement of DNA. In addition, these mitochondria contain the entire machinery required for the synthesis of DNA-dependent RNA. Although little is presently known about the biological function of the DNA contained in mitochondria, it seems interesting to follow the fate of mitochondrial information introduced into the zygote during fertilization. In mammalian sperm, the mitochondria are contained in the so-called midpiece. During fertilization, this midpiece, with its associated mitochondria, penetrates the egg.

After a short time, however, the mitochondria contained in the mid-piece disintegrate. This observation was made possible by careful electron microscopic examination of mammalian zygotes at successive times after sperm penetration. The fate of the mitochondrial DNA is not known, but presumably it also disintegrates. A much clearer situation exists in tunicates. The sperm of the tunicate *Ascidia nigra* contains only one mitochondrion, which is wrapped around the sperm nucleus. When the sperm first establishes contact with the external egg envelope (Fig. 5-5), it discards its only mitochondrion and then penetrates the egg. Clearly, the sperm does not contribute mitochondrial information to the zygote. Any traits caused by mitochondrial DNA would therefore have to be inherited maternally in tunicates, but since each egg contains many mitochondria, these may represent a genetically heterogeneous population.

## The operon concept in higher organisms

One of the strongest conceptual influences on developmental genetics has arisen from research in bacterial genetics. The DNA-RNA-protein dogma with its regulatory circuitry has brought about a major change in the thinking of developmental biologists. One aspect of this dogma, called the operon theory, is discussed in great detail in the volume *Gene Action.** Briefly, the operon is the unit of genetic transcription. It is composed of several parts (Fig. 5-6): structural genes, regulator genes, and an operator site, perhaps also an enhancer site. The structural genes, each encoding an enzyme of a biochemical pathway, are clustered adjacent to one another along the chromosome. This cluster of structural genes is coordinately controlled by the interaction of the product of a regulatory gene with the operator site located at one end of the cluster. The regulatory gene, on the other hand, may be located elsewhere in the genome, not adjacent to the other genes of the operon. It codes for a substance, the repressor, that inhibits operon function. Such a genetic system may be controlled by a small molecule, such as a substrate of the metabolic pathway. This substrate molecule, by combining with the repressor substance, prevents the repressor from binding to the operator and releases the structural genes to transcribe mRNA.

Can the operon concept be applied fruitfully to higher organisms, and is it significantly involved in differential gene function during cell differentiation? We propose to address ourselves to this question by first asking a number of additional more specific questions.

* P. E. Hartman and S. R. Suskind, *Gene Action* 2nd Ed. (Englewood Cliffs, N.J.: Prentice-Hall, Inc.), 1965.

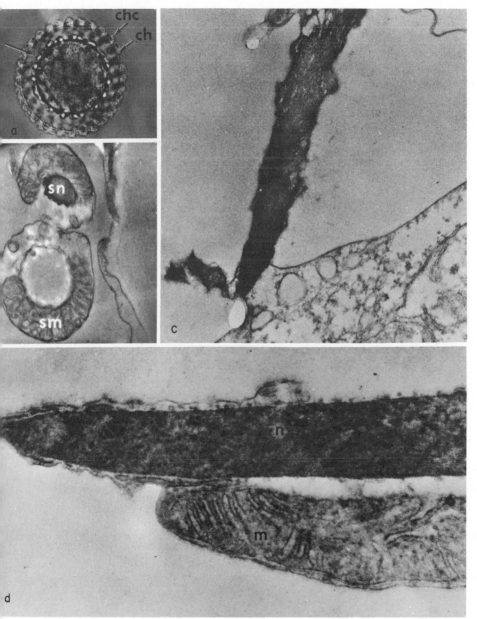

Fig. 5-5 The loss of the paternal mitochondrion during sperm entry into tunicate eggs. The eggs of tunicates are surrounded by various layers of cellular and non-cellular envelopes, shown in (a) t = test cells; ch = chorion; chc = chorion cells. A normal spermatozoan contains one large mitochondrion (m) parallel to its nucleus (n). Before penetrating through the chorion, the sperm discards its mitochondrion and completes fertilization in its absence. Notice in (b) cross-sections through two "spermatozoa," one a sperm with a nucleus (sn) and mitochondrion, the other a ghost, with just the mitochondrion (sm).

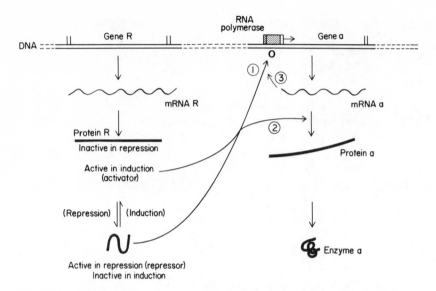

**Fig. 5-6** Structure and function of the operon. A general model for induction and repression of specific protein biosynthesis. The figure shows some of the major components that are presumed to take part in the regulation of genic activity and some possible sites of interaction by the components. A gene, R, is responsible for the synthesis of a protein that can exist in two conformations. In one conformation the protein is inactive as a repressor. In the second configuration (bottom) the protein has affinity for a chromosomal location, O (step 1). When the repressor protein is at O, RNA polymerase is unable to synthesize mRNA along gene a. When repressor protein is absent from O, RNA polymerase initiates synthesis of mRNA. A modification of this general scheme is diagrammed as reactions 2 and 3. In this modified model, the repressor acts at the level of mRNA causing it to remain complexed with the structural gene and to halt further mRNA synthesis at this place on the chromosome. The amount of the repressor available for inhibition of mRNA synthesis depends upon the rate of synthesis of the repressor protein, upon the repressor's affinity for O, and upon the presence of molecules of small molecular weight that influence the interconversion of the two forms of the repressor. In induction, the metabolite maintains the repressor molecule in a form that is inactive as a repressor. In repression, the metabolite enhances the conversion of repressor molecules from the inactive form to a conformation that is active as repressor.

## Are there examples, in higher organisms, for clustering of structural genes of biochemical pathways?

An example of gene clustering similar to that found in bacteria is represented in the histidine biosynthetic pathway in yeast, where the genes for three of the enzymes are clustered. Likewise, five genes of the aromatic amino acid pathway in *Neurospora* are closely linked. Higher on the evolutionary scale, a very good example of gene clustering has been described for the complex ribosomal RNA locus of *Xenopus*. As we have seen earlier, the nucleolar organizer region of the *Xenopus* chromosome contains the genes for 28 S and 18 S ribosomal RNA, each of which in a diploid genome exists in some 1000 copies. These genes are apparently arranged in alternating sequence (28 S–18 S–28 S–18 S–) along the chromosome, with some nonribosomal DNA stretches interposed between the ribosomal cistrons. The entire stretch of DNA is transcribed as a unit, and thus the 28 S and 18 S genes are synthesized coordinately. This may prove to be an unusual arrangement of genes in higher organisms. There are many clear-cut examples of lack of clustering of functionally related loci. The ribosomal RNA operon of *Xenopus*, for example, does not include the genes for ribosomal 5 S RNA, even though this RNA is transcribed coordinately with the 18 S and 28 S RNA. Likewise, the genes encoding the alpha and beta chains of the hemoglobin molecule are not linked in mammals. However, there are a few additional cases in which genes, apparently functionally related, are clustered. Perhaps the best analyzed case is the Ubx series in *Drosophila*, which will be discussed in a different context later (p. 175). This morphogenic pathway however has not yet been reduced to a biochemical pathway that might make it truly comparable to gene clustering in bacteria. This is also true for other morphogenic pathways, such as the genetically controlled tail malformations in the mouse, the genes for which are closely linked.

### Is there evidence for an operator in such gene clusters?

The gene clusters in *Neurospora* and yeast that were discussed previously are coordinately controlled. Thus, they are genetically functional units. The presence of an operator site associated with these clusters has not been established, since no mutations at an operator site have been found. This is also true for the ribosomal RNA operon in the frog, which clearly represents a functional unit, but in which an operator locus has not been demonstrated.

In any event, it is very doubtful that an operator locus is required for coordinate synthesis, and, in fact, clustering of genes is clearly not required as shown by the behavior of the cistrons for ribosomal 5 S RNA in *Xenopus*. Even in bacteria, cases have been reported in which enzymes appear coordinately controlled despite the fact that their structural loci are widely scattered throughout the genome. In higher organisms, many cases of synchronized increase and decrease of enzyme levels have been observed, and these are commonly interpreted to signify coordinate control at the genetic level. This interpretation must be regarded with caution for a variety of reasons. First, the level of an enzyme in higher organisms is controlled by at least three different mechanisms: rate of transcription, rate of translation, and rate of degradation. In higher organisms, transcription and translation are physically separated from one another, the former taking place in the nucleus, the latter in the cytoplasm. Thus, coordination could be brought about or disrupted at widely different positions in the cell. To speak confidently of coordinate transcription in the bacterial sense, we must first demonstrate the production, in the nucleus, of a polycystronic mRNA coding for several related enzymes, which has not been done. However, much recent evidence does indeed demonstrate that very high molecular weight RNA is synthesized in the nucleus. This RNA may, or may not, reach the cytoplasm, but if it does, it certainly does not retain its original size.

## Is there evidence for regulator genes?

By definition, regulatory genes must directly influence the productivity of a structural locus without affecting the quality of its product, and must furthermore be located outside the structural locus.

The example that superficially seems to fulfill some of these requirements is the xanthine dehydrogenase (XDH) system in *Drosophila* (Fig. 5-7). We choose this example because it is a case of "several genes– one enzyme." At least three genetic loci affect this enzyme: maroon-like (ma-l), located on the X chromosome; rosy (ry), located at locus 52 on the third chromosome; and low xanthine dehydrogenase (lxd), also on the third chromosome, but at locus 33, some 20 map units away from rosy. Flies that are genetically ry or ma-l are characterized by brown eye color.

It was observed early that ma-l and ry flies both lack XDH activity, but that lxd flies contain about 20% of wild-type XDH activity. For examination of the effect of these genes on the structure of XDH, the enzyme was purified and injected into rabbits to evoke the formation of antibodies against the enzyme. Later, serum from these rabbits contained antibodies that neutralized and precipitated xanthine dehydrogenase. Next, the mutant flies were investigated for the presence of cross-reacting material (CRM)—that is, material that can neutralize anti-XDH antibodies (Fig. 5-7). It was found that ry flies contained little if any CRM, which leads to the conclusion that ry is probably the structural locus responsible for the formation of XDH protein. Maroonlike flies, and lxd flies, on the other hand, were found to con-

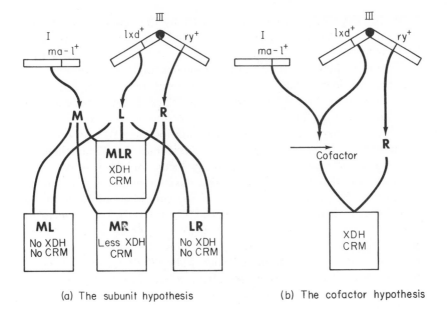

(a) The subunit hypothesis          (b) The cofactor hypothesis

Fig. 5-7   Models of the genetic control of xanthine dehydrogenase in *Drosophila*.

tain CRM in amounts equal to the quantity of XDH in wild-type flies. This result demonstrates that neither lxd nor ma-l are bona fide regulatory loci, since they do not affect the amount of enzyme protein produced at the structural locus. Nevertheless, lxd is interesting because it does affect the level of active XDH. In this sense, and because of its genetic map location outside the structural locus, it somewhat resembles a regulatory gene. Let us briefly describe how the analysis of this putative regulatory system has been pursued further.

*Is rosy a structural gene?* The evidence is indeed convincing. We have already mentioned that rosy flies do not contain CRM. Furthermore, the number of ry alleles in a fly determines its XDH content: ry/+ heterozygotes contain half as much XDH as +/+ wild-type flies, and flies of the genotype +/+/+ contain three times the haploid amount. By contrast, neither ma-l+ nor lxd+ loci show such a dosage effect. Finally, an electrophoretic variant was found that, on genetic mapping, was discovered to be at the ry locus.

*What is the role of ma-l and lxd?* Remember that maroonlike flies totally lack active XDH but contain a normal amount of CRM. The formation of active XDH apparently requires an interaction of materials produced by lxd and ma-l with the structural ry product.

That this interaction occurs is indicated by the observation of complementation. When extracts of ry and ma-1 flies are mixed, an active XDH is formed in vitro. Thus, ma-1 flies contain a ry complementing factor, and ry flies contain a ma-1 complementing factor. Interestingly enough, lxd flies contain a notable amount of ry complementing factor, but no ma-1 complementing factor. There is some evidence that the complementing factors are protein. We can now envisage at least two different mechanisms by which regulation of XDH by ma-1 and lxd might be brought about. First, XDH may be a multimer of polypeptide subunits encoded by the ry locus, and also by the other loci. One would have to assume in this model that the antigenicity of the finished XDH molecule resides in the subunits produced by the ry locus. Alternatively, one could assume that only ry produces an XDH subunit, and that ma-1 and lxd are responsible for the formation of enzymes involved in the production or attachment of XDH cofactors, such as flavinadeninedinucleotide (FAD), molybdenum, or iron. Both models are experimentally testable, but both are clearly different from bacterial regulatory schemes.

The precise nature of the regulatory action of the various genes controlling XDH is not understood. This is also true for many mutants that may also be regarded as regulatory in nature—that is, for the so-called suppressor mutants.

## Is there evidence for a repressor substance?

The theory that has recently attracted the greatest attention in the field of genetic regulation assumes that histones are repressor substances in the chromosomes. Histones are basic proteins that form ionic complexes with DNA when both are incubated together in vitro. Such DNA-histone complexes do not function as templates for RNA synthesis.

We shall discuss histone chemistry later. At this point, we wish to present the experimental evidence supporting the assumption that histones are repressor substances. In these experiments, RNA synthesis was measured in systems primed by DNA or chromatin. A typical reaction mixture contained radioactive ATP, GTP, CTP, and UTP, magnesium, mercaptoethanol, a buffer, RNA polymerase, and 25 µg of DNA. After such a mixture was incubated at 37°C, RNA was isolated from it and the radioactivity incorporated into RNA was measured. When deproteinized DNA is used in this experiment, active incorporation of precursor molecules into RNA is observed (Fig. 5-8). When

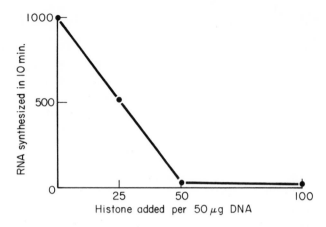

Fig. 5-8   Effect of histones on DNA-dependent RNA synthesis in vitro. (From R. C. Huang and J. Bonner. 1962. Proc. Nat. Acad. Sci. U.S. 48:1216. Fig. 3.)

25 μg of histone is added to the reaction mixture, the formation of RNA drops sharply; the addition of 50 μg of histone prevents all RNA synthesis. In this experiment, histone clearly prevents the transcription of RNA from a DNA template, and in this sense it is a repressor substance.

As we shall discuss in more detail later, only a few varieties of histones are identifiable in cells of higher organisms. Thus, it is difficult to envisage a mechanism by which histones could act as gene-specific repressors. Also, there is no evidence that histones originate at regulator loci as defined in the operon model. We should not forget, however, that RNA and nonhistone proteins are also associated with DNA in chromatin. Taken together, these various macromolecules might well serve as gene-specific regulatory substances. In fact, the molecular regulators of gene function must be part of the chromosome, and we may be confident that all the molecules found on chromosomes play some role in chromosome structure and function. Precise identification and characterization of chromosomal molecules and sorting out their various roles present one of the most important and exciting areas of contemporary research in developmental biology.

## Position-effect variegation

In the bacterial concept of the regulator gene, it is assumed that the regulator gene exerts its influence on the operator through a diffusible agent, the repressor substance. But this is not the only possibility. The mere position of a gene on a chromosome might influence the activity of adjacent or nearby genes. Indeed, there are cases supporting this idea. One of the mutations affecting eye pigmentation in *Drosophila* was described many years ago. Flies that carry this mutation have white eyes, rather than the dark red eyes typical of the wild type. The mutation is called white (w), and is located on the X chromosome, more precisely, in the euchromatic portion of the X chromosome (Fig. 5-9). Exposure of flies to X rays induces chromosome breaks. Broken ends of chromosomes have a tendency to stick to other broken ends, which sometimes leads to translocation of one part of a chromosome to another. In Fig. 5-9, such a translocation is diagrammed. In this case, a segment of the X chromosome with its white locus has been translocated to the fourth chromosome. Not only is the locus now on a new chromosome, but it is in close proximity to heterochromatin. Through appropriate genetic crosses, flies can be produced with a genetic constitution such as that shown in the lower part of Fig. 5-9. These flies have the genetic constitution R(w+)/w, where R stands for "rearrangement." From this genetic constitution, one would expect that the eyes of such flies would be wild type in pigmentation, because white is a recessive mutation. Instead the eyes are mottled or variegated; some cells of the eye are red, some white. This mottling is also observed in flies of the genotype R(w+)/R(w+), but not in flies of the genotype R(w+)/(w+) or R(w)/(w+). In other words only when the rearrangement is heterozygous for a mutant allele (w), or when the fly is homozygous for the rearrangement, does variegation occur.

Several cases of position-effect variegation have been reported, and a number of common denominators have been found. First, the recipient chromosome for the translocation is unimportant. The essential requirement is that the translocation be brought into the proximity of heterochromatin. Second, variegation is always associated with chromosome breakage, and at least one of the breaks leading to the translocation must lie within the heterochromatic portion of a chromosome. Thus, genes that normally reside in euchromatin exhibit variegation behavior in heterochromatin. The reverse is also true: when placed into a euchromatic environment, genes normally residing in heterochromatin frequently show an altered expression.

What is the mechanism underlying this peculiar behavior? At least two possible explanations are worth considering. In the case of the w variegation, the cells of the white area in the mottled eye may differ in their genotype from those of the red area as a result of somatic mutations induced by proximity to heterochromatin. Alternatively,

Chromosome constitution          Eye color

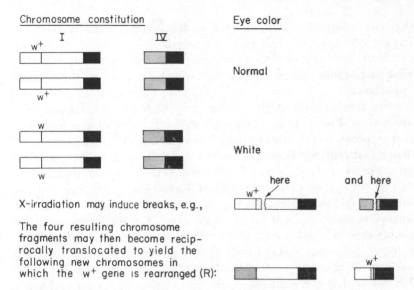

X-irradiation may induce breaks, e.g.,

The four resulting chromosome
fragments may then become recip-
rocally translocated to yield the
following new chromosomes in
which the w⁺ gene is rearranged (R):

Through appropriate crosses, flies with various chromosome constitutions
can be produced. For example:

Chromosome constitution          Symbol      Eye color

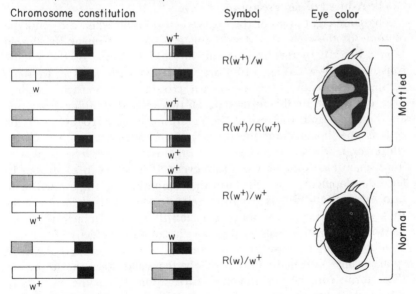

Open bars: euchromatin of chromosome I.
Black bars: heterochromatin
Grey bars: euchromatin of chromosome IV.

Fig. 5-9   Position effect variegation at the white locus in *Drosophila*.

the cells of both the white and red areas may be genetically identical, but the expression of their genotype may be influenced by the hetero-chromatic environment. All experimental evidence supports the latter view and argues against somatic mutation as a cause of the mottled appearance.

Since the behavior of these genes may be only a special case of a more general mechanism of chromosomal regulation of gene function, it may prove useful to examine in more detail the experimental evidence that led to this conclusion. First, let us look at some mottled eyes. The examples shown in Fig. 5-10b are taken from an investigation of cell lineage in the embryological formation of the eye and are directly applicable to our discussion of the position effect. These cell-lineage experiments will be discussed later. For the moment, notice simply that in mottled eyes, mottling is not a salt-and-pepper arrangement of differently colored cells or cell groups. Rather, all observed patterns fit precisely into a small number of eye sectors. Notice in Fig. 5-10b that the mottling occurs in positive and negative images, indicating some common pattern. Careful analysis of many mottled eyes, mottled because of position effect, shows that the different patterns can be reduced to eight sectors of the eye.

The same eight areas of the eye were also found when the genotype of some of the cells of a developing eye was experimentally altered (Fig. 5-10c, d). In this experiment, *Drosophila* larvae of the genetic constitution w/w^co (Fig. 5-10a) are irradiated with X rays. In some cells, this procedure leads to somatic crossing-over. Somatic crossing-over can result in the formation of two cells that differ from one another, and from the rest of the organism, in genotype. One of the cells is now homozygous for white and the other is homozygous for white coral, whereas the rest of the organism is heterozygous, w/w^co. This somatic crossing-over is a rare event. If it occurs in a cell that in later development is to help form the composite eye, the eye will contain this cell and all its daughter cells. In the eye that forms under these conditions, a spot of mutant tissue will be seen. In fact, two spots, or twin spots, will be seen in this case, because the progeny of the pair of mutant cells grow side-by-side. In Fig. 5-10, we see that the twin spots formed fall again into precisely the same areas in the eye as those formed in the position-effect variegation. When the outlines of all mottled eyes are traced on one eye drawing, a distinct clustering becomes apparent. Since somatic crossing-over is a rare event, we can safely say that the spots represent cell clones derived from a single pair of cells in which the mutational event occurred. Cell lineage assures the propagation, through cell heredity, of the mutant trait. Since position-effect sectors fall into the same pattern, we can conclude that these

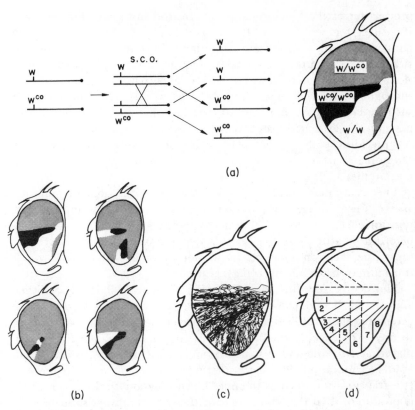

Fig. 5-10 Somatic crossing-over and the production of mottled eyes in *Drosophila*. (a) A larva of the genetic constitution $w/w^{co}$ is x-rayed. This can lead to somatic crossing-over (s. c. o.). In the ensuing cell division, two genetically different cells are produced: one is of the constitution $w/w$, the other, $w^{co}/w^{co}$. The cell clones derived from these two cells later differentiate into two adjacent groups of eye cells differing in color because of their genotype, and appear on the background of that portion of the eye where the stem cells show no somatic crossing over ($w/w^{co}$). (b) Four different mottling patterns among the many that have been observed. (c) Outlines of many mottling patterns. (d) The eight postulated cell clones, responsible for the patterns. (From H. J. Becker. 1956. Verhandl. Deutsch. Zoolog. Ges., p. 258.)

sectors, too, are made up of clones of cells that were initiated by a rare event, either somatic mutation or a heritable alteration of gene expression.

Several lines of evidence argue against somatic mutation as the cause of variegation. First, when flies of a genetic constitution that leads to variegation are raised at low temperature, the amount of white tissue in the eye is greatly increased. If variegation were caused by a somatic mutation, this would mean that low temperature favored the production of the somatic mutation. Instead mutations increase with increasing temperature, not the reverse. The observed influence of low temperature on the extent of variegation is in conflict with the assumption of somatic mutation. Second, if somatic mutation initiated variegation, then such mutations would be expected to occur randomly throughout the body, including the cells in the germ line; however, such mutations have not been observed. Third, the biochemical composition of variegated eyes reveals excessive amounts of substances of the pteridine pathway that play an important role in the formation of eye pigments in wild-type organisms. These substances are absent in mutant flies.

The accumulated evidence argues strongly against somatic mutation as a cause of variegation, and we are left with the assumption that position-effect reflects an inherited alteration of gene function. The heterochromatin, according to this hypothesis, affects the activity of the white locus in some cells and not in others, and this effect is transmitted through many cell generations.

The degree to which this alteration of gene function is achieved appears to be related to the distance of the affected locus from heterochromatin, as demonstrated whenever the spreading effect occurs. In these experiments, the behavior of two genetic loci, *white* and *split,* placed at two different distances from heterochromatin, is compared. Split is a mutation that affects the arrangement of facets in the eye. If a translocation, such as the one shown diagramatically in Fig. 5-11, is produced, then the white areas of the eye are always split, but the converse is not true. Some split areas are not white. In other words, the split phenotype may be altered alone, but altered white phenotypes occur only in combination with altered split phenotypes. This polarity in expression argues for a spreading effect originating in the heterochromatin and moving toward the euchromatic genes. If it were possible to create a chromosome in which split and white were at equal distances from heterochromatin, then one would expect the polarity to break down and indeed, it does. If the euchromatic region containing split and white is inserted into heterochromatin (Fig. 5-11), the resulting eyes show no spreading effect. In this arrangement, white may or may not be combined with split, and split may or may not be combined with white.

Position-effects therefore represent cases in which the activity of a gene is influenced by its position with respect to other components of the genome. The precise mechanism by which this gene-gene interaction is brought about is not understood, nor is it understood why some cells show the altered phenotype whereas others do not.

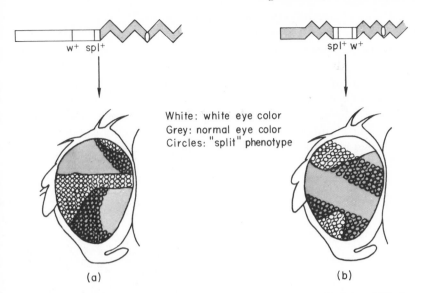

White: white eye color
Grey: normal eye color
Circles: "split" phenotype

(a)  (b)

Fig. 5-11   The spreading effect of proximity to heterochromatin. (From W. Baker. 1963. Am. Zool. 3:66. Fig. 5a.)

## The Lyon hypothesis

In the preceding section, the effects of chromosomal location on individual gene expression were discussed. In this section, we take up a more dramatic example of intragenomic effects on gene expression—in fact, on the expression of the entire X chromosome. In mammals, the female contains two X chromosomes, the male an X and a Y chromosome. Thus, all genes on the X chromosome are present in duplicate in the female and singly in the male. Nevertheless, those genes known to be on the X chromosome—that is, the sex-linked genes —appear to function equally well in both female and male. A suggestion made years ago is that a mechanism for dosage compensation must exist in the female so that the double dose of the X chromosome genes is functionally equivalent to the single dose of these same genes in the male. The original explanation postulated the existence of modifying genes; this explanation was logically adequate but not very persuasive. Moreover, evidence for it was slight.

Recently, a new explanation for the behavior of the X chromosome of mammals was put forward on the basis of facts from investigations in mouse genetics and physiology. This explanation has general significance for our understanding of the regulation of gene and chromosomal functions during the differentiation of mammalian cells.

It had been observed by Barr that female cells usually contain a clump of heteropyknotic chromatin (Fig. 5-12). This body, now called the Barr body, is absent from the cells of the male. Other investigators showed that the cells of female mice contained one chromosome that was heteropyknotic, and suggested that the so-called sex chromatin or the Barr body was, in fact, the heteropyknotic X chromosome.

With this suggestion in mind, certain genetic facts were noted. First, mice of the chromosome type XO are normal fertile females, thus proving that only one active X chromosome is necessary for normal de-

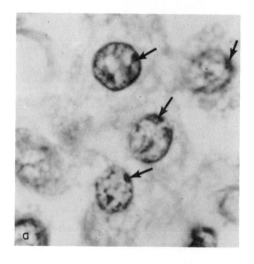

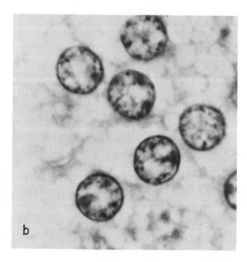

Fig. 5-12 (a) A photograph of a human female cell showing the Barr body. (b) Photograph of a male cell without the Barr body. Arrows indicate Barr bodies. (Courtesy of M. L. Barr.)

velopment of the female mouse. A second important genetic fact stems from observations on a variety of coat color mutants in the mouse. These mutants, variously described as brindled, mottled, tortoise-shell, dappled, and so forth, are all lethal in the male and viable only in heterozygous form in the female. Moreover, as the names suggest, these pigment patterns are mixtures of differently colored hairs.

An explanation for sex chromatin and for the genetic behavior of these genes was formulated in the following hypothesis. There are three major parts. First, the heteropyknotic X chromosome is geneti-cally inactive. Second, the active chromosome can be derived from either parent. Third, the inactivation of one of the X chromosomes occurs during early embryonic development and persists thereafter in all the descendant cells. The hypothesis, sometimes briefly referred to as the "single active X" hypothesis, provides an adequate explanation for the Barr body and also explains many pigment patterns in mice. The pigment cells in female mice would contain only one active X chromosome. If the melanocyte contained an active X, bearing genes unable to support the synthesis of pigment, then that melanocyte would remain colorless. On the other hand, if the melanocyte contained the active chromosome carrying normal genes for pigment, then the melanocyte would synthesize melanin. Such pigment cell genotypes lead to a pigment mosaic reflecting the distribution of pigment cells with different active X chromosomes.

A uniquely satisfying test of the Lyon hypothesis as it is also called, has been provided at the enzyme level. Use was made of the fact that the enzyme glucose-6-phosphate dehydrogenase (G-6-PD) is encoded in a gene on the X chromosome. Moreover, this gene exists in two alternate allelic forms, each coding for an enzyme molecule with different electro-phoretic properties. Thus, with conventional isozymic techniques pre-viously discussed with reference to LDH, it is possible to resolve in starch gels tissue homogenates containing G-6-PD so that two recogniza-ble areas of enzyme activity are revealed. A female heterozygous for this gene produces both forms of the enzyme. Males, however, can produce only one form, since they are hemizygous and never possess both alleles of the G-6-PD gene simultaneously. It was reasoned that if the Lyon hypothesis applied, then some of the cells of the female would be producing one form of the enzyme and the other cells the alternate form, but no cell would produce both forms. To test this, small pieces of skin from human females known to be heterozygous for the G-6-PD gene were isolated by biopsy. The cells were cultured in vitro and then separated and cloned. A zymogramic analysis was then made of the different cloned cells. The results are remarkably clear: each clone exhibited only one of the two forms of the enzyme, but not both

(Fig. 5-13). Thus, each cell giving rise to a clone contained only a single active X chromosome. These results strongly support the Lyon hypothesis.

However, the presence of only one form of G-6-PD in any particular cell does not imply that all genes on the inactivated X chromosome were totally nonfunctional. In fact, there is some evidence to suggest that the inactivated X chromosome is not totally inactive. For example, if one X were sufficient, then XO human females should be completely normal; but they are, in fact, somewhat abnormal. In mice, autosomal genes translocated to one of the X chromosomes are not always completely inactivated. Nevertheless, all supernumerary X chromosomes (more than one) are certainly largely inactivated, which makes possible the existence of human females with as many as four X chromosomes. No other chromosome in the genome can exist in aneuploid multiplicity without generating extreme abnormalities of development and

Fig. 5-13   Zymogram analyses of G-6-PD from homozygous and heterozygous human female cells (after Davidson et al. PNAS 50: (1963) 484, fig. 1). Note that each clone contains only one isozymic form of G-6-PD. The particular isozyme synthesized depends upon the random inactivation of one of the two X chromosomes, each carrying an allelic gene for one of these two isozymes.

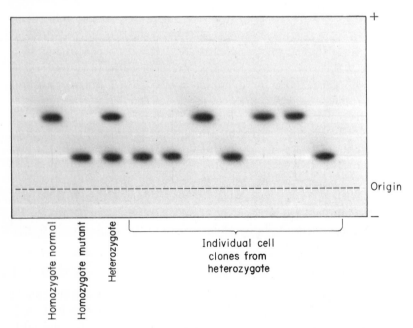

function. Clearly X chromosomes are subject to special regulatory controls not applicable to other chromosomes, but a cytological consequence of inactivation is that the inactivated X becomes heterochromatic. Such heterochromatin occurs on a limited scale in all chromosomes and may be due to similar chromosomal changes, although on a reduced scale compared with the X.

## Heterochromatin

Heterochromatin has attracted the attention of cytologists since 1928, when it was first described in the mouse. Heterochromatin is that chromatin which remains compacted throughout interphase of the cell cycle; the remainder of the chromatin—that is, euchromatin—becomes dispersed. These two types of chromatin stain very differently, heterochromatin very deeply and euchromatin practically not at all. Probably all chromatin can exist in either the heterochromatic or euchromatic configuration, in accord with its functional state. Heterochromatic regions of the chromosomes are judged to be inactive, as indicated by several observations. First, active genes are seldom mapped in heterochromatic regions. The heterochromatic X chromosome in mammalian females is inferred to be almost completely inactive. Certainly the ability to incorporate tritiated nucleotides into RNA is much reduced. Moreover, the paternal chromosome complement in male mealy bugs is genetically inactive (Fig. 5-14), and these chromosomes are heterochromatic. Thus, transcription of RNA is not characteristic of heterochromatin, as it clearly is of euchromatin. This is

Fig. 5-14 Heterochromatized male genome of the mealy bug.

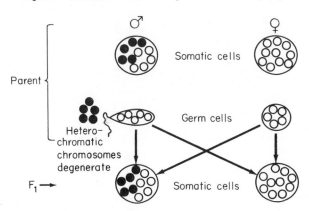

true for restricted heterochromatic regions of chromosomes, for single chromosomes—such as the inactivated X chromosome of female cells, and for entire chromosome sets—such as the male genome in the mealy bug. Thus, heterochromatization is uniformly correlated with the cessation of RNA transcription (Fig. 5-15).

Nevertheless, heterochromatin replicates, as demonstrated by the incorporation of DNA precursors, but the replication is retarded when compared to euchromatin. These observations have led to the suggestion that heterochromatin and euchromatin are alternative interchangeable physical states of the chromatin. If this is true, then differential gene inactivation may simply be a highly localized expression of the more general phenomenon of heterochromatization, as seen in large areas of single chromosomes, in whole chromosomes (one of the X chromosomes in mammals), and in entire chromosome sets, as in the mealy bug. The molecular mechanisms underlying this heterochromatization would then control the degree of condensation or coiling of the chromosome to render the DNA strands inaccessible for RNA transcription and to retard the replication of the DNA itself.

Fig. 5-15   Photograph showing heterochromatin in inactive erythrocyte nucleus (a) after fusion with cell containing an active nucleus (b) with mostly euchromatin. (Courtesy of H. Harris.)

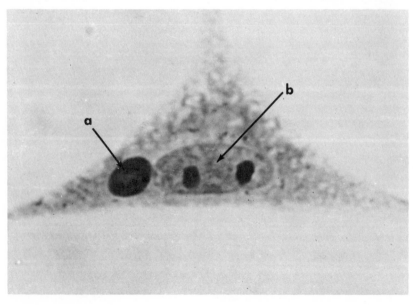

## Hormones and gene action

Hormones have long been known to stimulate enzyme activity. Recently, elegant experimentation has shown that in some instances the increased enzyme activity is due to increased enzyme synthesis. Activation of inhibited enzyme molecules without additional synthesis also leads to increased activity. This is a common metabolic device for regulating enzyme activity. The work on enzyme synthesis referred to above was done on the enzyme tryptophan pyrrolase in rat liver. This enzyme breaks open the indole ring of tryptophan and leads to the synthesis of an essential vitamin, nicotinamide.

The experiments to be described were performed on adult rats. Every 4 hours, hydrocortisone, tryptophan, or a combination of the two were injected intraperitoneally. And every 4 hours some of the animals in the experimental series were killed and enzyme activities in their livers determined. As Fig. 5-16 shows, tryptophan and hydrocortisone each leads to increased enzyme activity, and a combination of the two substances leads to an increase that is far greater than would be true if the increase were simply additive. Notice also that the kinetics of enzyme increase induced by the two substances differ greatly. This observation led the investigators of this system to speculate that the two compounds induced the increase by entirely different mechanisms. More specifically, it was hypothesized that hydrocortisone induced synthesis of enzyme, whereas tryptophan prevented its degradation.

Further experiments showed that this hypothesis was essentially correct. Crucial support came from a series of experiments in which radioactive labeling of tryptophan pyrrolase and specific antibody precipitation techniques were combined. In one experiment, the enzyme was first purified and then injected into rabbits so as to produce antibodies against the enzyme. With such antibodies, it was possible to answer the specific question whether tryptophan pyrrolase was synthesized in response to the injected hormone or substrate or whether it was merely being released from an inhibited state. Either saline, or hydrocortisone, or tryptophan was injected into rats again, and 3 hr later radioactive amino acids were also administered. Still later, the livers of these animals were removed and homogenized, and, as one would expect, all newly synthesized proteins were found to be radioactive. Newly synthesized tryptophan pyrrolase was now singled out from among the rest of the proteins by precipitation with the specific antibody, which was added to the liver homogenate. Figure 5-17 presents the results of such an experiment. Notice that new

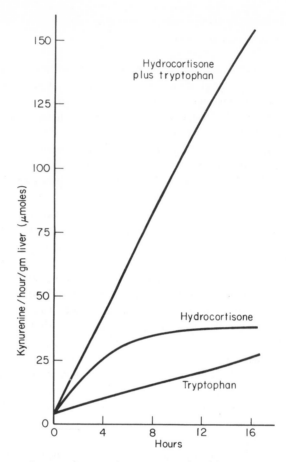

Fig. 5-16 Increase in tryptophan pyrrolase activity produced by repeated administration of hydrocortisones and/or tryptophan. Injections were given every four hours. (From R. T. Schimke. 1967. Nat. Cancer Inst. Monograph **27**:301–314. Fig. 4.)

tryptophan pyrrolase molecules were made in response to hydrocortisone but not in response to tryptophan. We must conclude, therefore, that hydrocortisone induced enzyme synthesis.

The same type of experiment was also used to examine the possibility that tryptophan acts to stabilize the enzyme once it has been made (Fig. 5-18). Again, proteins in the living rat were labeled with radioactive amino acids. At some later time, either saline or tryptophan was injected. Still later, the livers were homogenized and the amount of radioactive tryptophan pyrrolase still present was determined by precipitation with antibodies. As shown in the figure, the amount of

| Injected material | Enzyme activity after injection (units) | Enzyme protein radioactivity after injection (counts/min.) |
|---|---|---|
| NaCl | 41 | 1406 |
| Hydrocortisone | 189 | 9466 |
| Tryptophan | 80 | 1954 |

Fig. 5-17  Induction of tryptophan pyrrolase synthesis by hydrocortisones, and lack of induction by tryptophan.

labeled tryptophan pyrrolase declines steadily after saline injection, but remains constant in the presence of tryptophan.

In an earlier section of this chapter, we raised the question whether operons, in the bacterial sense, existed in higher organisms. We described a few examples in which one or another component of a bacterial operon was manifested in a higher organism—for example, the physical clustering of related loci on chromosomes, or the existence, operationally, of repressor substances. Perhaps hormones should be added to this list, since they may play a role analogous to bacterial effector molecules. We should caution, however, that the molecular mechanisms underlying hormone action in the genetic regulatory machinery are quite unknown. Indeed, the physical and/or metabolic distance between hormone and gene may be very great.

In the discussion of the effect of hydrocortisone on tryptophan pyrrolase, we have learned that the hormone leads to elevated enzyme

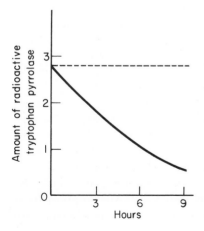

Fig. 5-18  Stabilization of tryptophan pyrrolase by tryptophan. Prior to this experiment, tryptophan pyrrolase of rats was made radioactive by injecting radioactive amino acids. Then, at zero time on the abscissa, and also after four and eight hours, the animals received injections either of saline (————) or of tryptophan (_____). At the times indicated on the abscissa the amount of radioactive enzyme was determined. (From Schimke. Fig. 3.)

synthesis. In this sense, it regulates gene activity. This is by no means an isolated observation. A large number of other liver enzymes have been found to increase after administration of cortisone. Other hormones have similar effects. Thyroxine controls the transition from larval to adult hemoglobin in amphibian metamorphosis. Protein synthesis in the uterus of mammals is under the control of estrogens, and other organs show increased protein synthesis after administration of androgens.

In many of these cases, it has been shown experimentally that the hormones induce enzyme synthesis and do not merely activate already synthesized molecules. The experimental design of these experiments was similar to the one discussed earlier for tryptophan pyrrolase. Alternatively, metabolic inhibitors may be used. Actinomycin, which prevents the DNA-dependent synthesis of RNA, and also puromycin, which inhibits protein synthesis at the translational level, both prevent hormone-mediated increases in enzyme activity, indicating that protein synthesis is involved. The results of recent research increasingly imply that hormones act by derepressing the genetic material either by interacting directly with chromatin or through a sequence of intermediate steps.

That de novo protein synthesis is, indeed, induced by hormonal treatment is perhaps best illustrated by the increase of α-amylase activity in barley endosperm during seed germination. The embryo of a germinating barley seed releases gibberellic acid to the aleurone layer, which consists of living cells surrounding the nonliving endosperm. These cells respond to the hormone by producing the enzyme α-amylase, which then passes into the endosperm and digests the starch. When barley embryos are separated mechanically from the endosperm and aleurone, the latter will not produce amylase unless gibberellic acid is added. The α-amylase that is formed after hormonal treatment will become labeled if radioactive amino acids are also provided. The newly formed, labeled enzyme has been demonstrated to be identical to the α-amylase synthesized during normal development. "Fingerprinting" of tryptic digests of the induced enzyme showed convincingly that it was indeed synthesized de novo.

The mechanisms by which enzyme synthesis is induced by hormones have been extensively investigated, but without decisive conclusions. One hypothesis assumes the existence in target tissues of receptor sites that recognize and bind the hormones. This assumption is supported by the observation that injected tritium-labeled estradiol, for example, is preferentially localized in the uterus of immature female rats, but not in other tissues. The hypothesis has received further support from observations that binding of a given hormone is affected by the concentration of competing hormones. If, for example, tritiated estradiol is injected in the presence of diethylstilbesterol, a synthetic estrogen, the amount of estradiol bound to uterine tissue decreases with increased concentration of the injected competitor. From this type of competition experiment it was calculated that approximately 2500 binding sites for estradiol exist in every cell of the uterus.

Much recent work has been devoted to the exploration of the chemical nature of these binding sites. One set of experiments consisted in fractionating the uterine cell after injection of radioactive estradiol, and investigating which of the subcellular fractions contained the labeled hormone. More than half of the hormone could be recovered from a heavy subcellular fraction containing nuclei and myofibrils, and about 30% of the hormone was recoverable from the supernatant after a high-speed centrifugation (105,000g). When this soluble material is subjected to sucrose density-gradient centrifugation, the hormone sediments with a sedimentation value of about 9.5 S. This fraction was obtained from uterine tissue, but not from other organs, such as the intestine. The cell fraction bound to the hormone appears to be protein, since the hormone can be released from it by treatment with proteolytic enzymes, but not with ribonuclease or deoxyribonuclease. From sedimentation data, it appears to have a molecular weight of approximately 200,000. This same molecular species may also account for the observed hormone binding by the insoluble fraction—that is, by nuclei and microfibrils.

These studies traced the fate of estradiols after injection into living animals. In other experiments, organ cultures were used for identifying the site of hormone attachment. Endometrium, removed from the uterine walls of calves, was maintained in culture in the presence of radioactive estradiol. After this incubation, various subcellular components were isolated and their estradiol content ascertained. The results showed that the hormone was concentrated within the nuclei of cells, where it was found to be associated with chromatin.

All these experimental observations show hormones to be very interesting as potential gene regulators. The weight of evidence clearly indicates that hormones can induce protein synthesis, although perhaps indirectly. Hormonal effects can be achieved by a variety of mechanisms. First, they may stimulate mRNA synthesis by acting as gene derepressors. Indeed, it has been known for several years that the rate of RNA synthesis in rat liver, for example, increases after the injection of hydrocortisone. The same observation was also made in rat uterine endometrium after injection of estrogens, and in the dormant buds of potato tubers after administration of gibberellic acid. In the latter case, an increased rate of RNA synthesis per unit of DNA was demonstrated; this result is consistent with increased template activity. The increase could also be brought about by an entirely different mechanism, namely, activation of RNA polymerase, without additional templates. It would be helpful to know whether the hormone action leads to a locus-specific increase in gene action—that is, a locus-specific increase in mRNA synthesis. But we have pointed out repeatedly that the identification of a specific kind of mRNA is difficult. Here again, we must presently rely on observations of specific enzyme synthesis. There is little doubt that cortisone, for example, selectively induces the synthesis of a few enzymes, because the increase in specific radioactivity of these enzymes per unit tissue is far greater than that of total protein. But the enzyme specificity of hormone action has often been overestimated. This is again best seen in the case of rat liver enzymes. Figure 5-19 shows the increase in specific activities of four rat liver enzymes during cortisone administration.

Notice that tryptophan pyrrolase and tyrosine transaminase increase at a far greater rate than do glutamic-alanine transaminase and arginase. At first sight, it appears that cortisone stimulates selectively the first three enzymes while affecting the fourth only slightly, if at all. Careful experimentation has shown, however, that the four enzymes differ greatly in stability (Fig. 5-20). Accordingly, since these enzyme levels remain constant at steady state, their relative rates of synthesis even in normal livers must be very different. Nevertheless, the increase in rate of

Fig. 5-19    Increase of four enzymes in rat liver after administration of hydrocortisone. (From Schimke. Fig. 2.)

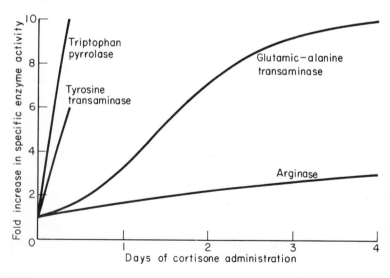

| Enzyme | Half-life (hr) | Basal enzyme activity (units) | Enzyme synthesized Basal (units/hr) | Cortisone (units/hr) | Ratio cortisone: normal |
|---|---|---|---|---|---|
| Tryptophan pyrrolase | 2.5 | 3.0 | .84 | 3.4 | 4.0 |
| Tyrosine-glutamic transaminase | 2.0 | 78 | 27 | 114 | 4.2 |
| Glutamic-alanine transaminase | 84 | 252 | 2.0 | 14.4 | 7.2 |
| Arginase | 96 | 13,800 | 138 | 534 | 3.9 |

**Fig. 5-20** Half-life and response to cortisone treatment of four liver enzymes. (From Schimke. Table I.)

synthesis produced by cortisone is also identical for all four enzymes, as is readily seen in the last column of Fig. 5-20. Clearly, the curves shown in Fig. 5-19 are deceptive. One could speculate at this point that the constant value of approximately 4.0 (Fig. 5-20) indicates coordinate control of the synthesis of these four related enzymes. Since increased rates of RNA synthesis are generally observed after cortisone treatment, the simultaneous change in activity of various enzymes might be brought about by coordinate control operating at the transcriptional level of gene function. However, no evidence is now available to identify the newly formed RNA with messenger RNA coding for the synthesis of these enzymes. In fact, the weight of the evidence thus far indicates that RNA of all major classes is formed in response to hormone stimulation, including DNA-like RNA, tRNA, and rRNA.

In addition to stimulation of transcription, hormones may also influence enzyme activity by controlling translation. Recent experimental results do lend some support to the view that hormones might, at least as secondary effects, exert control at the translational level.

Hormones, then, are reasonable candidates for a role in gene regulation analogous to the effectors in the bacterial regulatory system. However, they are organ specific rather than simply gene specific in effect. They do influence the synthesis of specific enzymes but not singly. Rather, groups of seemingly unrelated enzymes are affected together. Most significantly, hormones appear to become localized in the nucleus, more precisely, on chromatin. In our discussion of chromosome structure and function, we shall encounter hormones again as they affect chromosomal gene activity.

## Chromosomal macromolecules

Most of the genetic information of cells, encoded in DNA, is embedded in complex organelles, the chromosomes. Chromosomes contain, in addition to DNA, several classes of protein and some RNA. More information relevant to chromosomal structure will be presented later (Chap. 6). At this point, we wish to discuss the regulatory aspects of chromosomal macromolecules, particularly on interphase chromosomes, which are the metabolically active chromosomes of the cell. The chemistry of chromatin must be comprehended if the structure and function of chromosomes is to be understood. Chromatin is that fraction of a cell remaining after removal of tissue fragments, membranes, and other nonnuclear components. It can be sedimented readily by centrifugation and forms a clear, gelatinous pellet, consisting primarily of protein and DNA (in a ratio of about 2:1) and some RNA (about two-tenths the amount of DNA). Several ill-defined proteins make up the proteinaceous component of chromatin. In this fraction is also RNA polymerase, which forms an integral part of the chromosome and is necessary for genetic transcription. Isolated chromatin can be used for RNA synthesis in vitro without the addition of extra RNA polymerase, although with a very low yield of RNA. Generally, when chromatin is used for the study of DNA template activity in vitro, RNA polymerase is added to the preparation.

Also present in chromatin are histones, the basic groups of which are thought to be complexed through salt linkages with the phosphate groups of DNA. If one assumes that each phosphate group of DNA is matched by a basic group of a histone molecule, then a mass ratio of 1.35:1 (histone:DNA) would be required; this value was calculated from the amino acid composition of histones. In reality, however, this does not appear to be the case. Figure 5-21 shows the chemical composition of various chromatins, and it is apparent that the mass ratio varies considerably, indicating that DNA cannot be fully complexed with histones in most tissues.

Chemically, histones appear to be rather homogeneous proteins. When they are extracted from chromatin of diverse origins, such as from pea plants and calf thymus tissue, very similar fractions are found. The histone profiles shown in Fig. 5-22 are obtained when such preparations are chromatographed on ion exchange columns. When these major histone fractions are analyzed electrophoretically, each proves to be somewhat heterogeneous. By preparative electrophoretic techniques, rather pure histone fractions can be isolated, as judged by carboxyterminal and aminoterminal analyses. According to these

| Source of chromatin | Content, relative to DNA, of | | | | Template activity (% of DNA) |
|---|---|---|---|---|---|
| | DNA | Histone | Nonhistone protein | RNA | |
| Pea embryonic axis | 1.00 | 1.03 | 0.29 | 0.26 | 12 |
| Pea vegetative bud | 1.00 | 1.30 | .10 | .11 | 6 |
| Pea growing cotyledon | 1.00 | 0.76 | .36 | .13 | 32 |
| Rat liver | 1.00 | 1.00 | 0.67 | .043 | 20 |
| Rat ascites tumor | 1.00 | 1.16 | 1.00 | .13 | 10 |
| Human HeLa cells | 1.00 | 1.02 | 0.71 | .09 | 10 |
| Cow thymus | 1.00 | 1.14 | .33 | .007 | 15 |
| Sea urchin blastula | 1.00 | 1.04 | 0.48 | .039 | 10 |
| Sea urchin pluteus | 1.00 | 0.86 | 1.04 | .078 | 20 |

Fig. 5-21 Chemical compositions of chromatins. (From J. Bonner, et al. 1968. Science 159:47–56. Copyright 1968 by the American Association for the Advancement of Science.)

various tests, the histone species from pea plants and calf thymus are remarkably similar, even in amino acid composition.

Therefore, as genetic regulators, clearly histones alone cannot have the specificity required for locus-selective repression of genetic activity. Nevertheless, histones do generally repress DNA-dependent RNA synthesis. The removal of histones from chromatin is paralleled by an increased template activity, as expressed by RNA synthesis.

If histones, in addition to this quantitative control over template activity, are to exert a locus-specific action, then some modification of the molecule must impart specificity. In fact, chemical modification in the form of acetyl groups has been shown to occur in some histones and not in others. It is too early to say what the biological function of these chemical groups may be. Another molecule, chromosomal RNA, is also associated with histones and may be involved in specific interactions with DNA. RNA has been found associated with chromatin in pea plants, rat ascites tumor cells, rat liver, and calf thymus cells. The RNA may be bound to DNA by base pairing, which is indicated by the fact that chromatin-bound RNA is insensitive to ribonuclease unless deoxyribonuclease is also applied. As an RNA molecule, this chromosomal RNA is peculiar in that it contains large amounts of 5-methyl, dihydrocytidylic acid. It is covalently bound to chromosomal protein, and this complex is bound to histone by hydrogen bonds.

Chromosomal RNA hybridizes with DNA very slowly, indicating that it is rather heterogeneous. In competition experiments, it was found that transfer RNA and ribosomal RNA do not effectively compete with

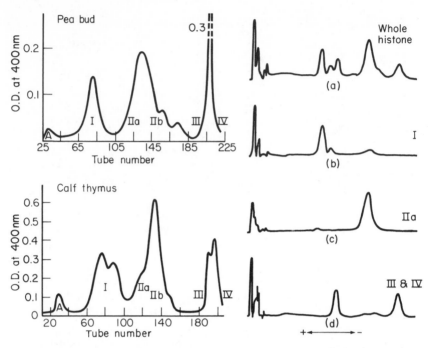

Fig. 5-22  Fractionation of histones by column chromatography (left) and by disc electrophoresis (right). Column chromatographic separation of histones from different sources, peabud and calf thymus, reveals a similarity in the major populations of histone molecules. Each of the classes of histone molecules can be further purified by disc electrophoresis as shown in the sequence of densitometric tracings shown on the right. At the top right is a preparation of whole histones resolved by electrophoresis. The lower three tracings on the right present the results of electrophoretic resolution of individual classes of histone molecules previously separated by chromatography. By these procedures, and others, a small number of essentially homogeneous protein preparations can be obtained. (From Bonner.)

chromosomal RNA for binding to DNA. Some evidence from hybridization experiments indicates that chromosomal RNA is also tissue specific. This supports the interesting notion that chromosomal RNA serves to recognize the DNA template and thus directs histones to specific sequences on DNA.

One should not overlook the possibility that nonhistone proteins play significant roles in regulating chromosomal function. As Fig. 5-23 shows, these proteins occur in great quantity and in different amounts in different chromatins. Histochemical evidence shows (p. 98) that active loci (puffs) in giant chromosomes contain more nonhistone protein than silent loci, whereas the amount of histone in a puff does not appear to change during transcription. Furthermore, elevated levels of nonhistone proteins have been found in regenerating liver cells (as compared with nonregenerating liver cells) and in several varieties of tumor cells. Also, chromatin prepared from different developmental stages of sea urchins differs widely in its relative content of nonhistone proteins: blastula chromatin contains little nonhistone protein, whereas chromatin obtained from plutei contains considerably more. Blastula chromatin is relatively inactive with regard to genetic transcription, whereas pluteus chromatin is very active.

Some progress is now being made in the study of these nonhistone proteins, in an attempt to ascertain their roles in the regulation of gene function in higher organisms. These proteins comprise at least 50% of the total nuclear proteins of a cell. Ion exchange chromatography and gel electrophoresis have revealed the existence of at least 20 different protein fractions in preparations of acidic proteins (nonhistone) from given tissues (Fig. 5-23). But it is difficult at this time to compare patterns obtained from different tissues because different extraction procedures have been used in the different investigations. The analysis of acidic proteins has been difficult because of their low solubility. Drastic measures have often been taken to solubilize these proteins, and it is therefore difficult to evaluate the results with reference to native configuration and tissue-specific patterns.

On the other hand, evidence is accumulating to show that these acidic proteins interact with histones, thereby affecting the repressor function of histones. If, for example, a histone preparation with a known inhibitory effect on DNA-dependent RNA synthesis is preincubated with phosphoproteins, its inhibitory effect is reduced. Operationally, the nonhistone component is thus acting as a dere-

Fig. 5-23 Non-histone proteins in chromosomes. Acrylamide disc electrophoresis of non-histone proteins from rat liver nuclei. Nuclear extracts were electrophoresed on a column of acrylamide gel, and the gel stained for the presence of protein. Nearly 20 different electrophoretic fractions are visible. (From W. Benjamin and A. Gellhorn. 1968. Proc. Nat. Acad. Sci. U.S. **59:**262–268. Fig. 2.)

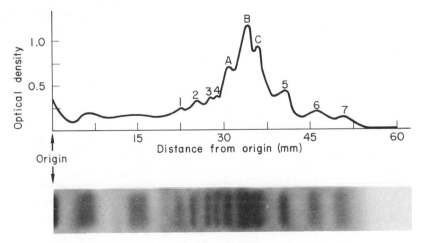

pressor substance. In line with this interpretation is the observation that in many interphase cells, the turnover of nonhistone proteins is greater than that of histones. In a study using ascites tumor cells in culture, and also rat liver in vivo, it was found that the labeling of DNA-associated nonhistone proteins was proportional to that of DNA-like RNA. Such results support the idea that nonhistone proteins are involved in the control of DNA-dependent RNA synthesis—that is, in the regulation of gene function.

## REFERENCES

Baker, W. 1963. Genetic control of pigment differentiation in somatic cells. Am. Zool. 3:57–70.

Becker, H. J. 1966. Genetic and variegation mosaics in the eye of *Drosophila*. *In* A. A. Moscona and A. Monroy [eds.] Current topics in developmental biology, vol. I, pp. 154–172. Academic Press, New York.

Bonner, J. and P. Ts'o [ed.]. 1964. The nucleohistones. Holden-Day, San Francisco. 398 pp.

Britten, R. J. and D. E. Kohne. 1968. Repeated sequences in DNA. Science 161:529–540.

Brown, S. W. 1966. Heterochromatin. Science 151:417–425.

Busch, H. 1965. Histones and other nuclear proteins. Academic Press, New York. 266 pp.

Callan, H. G. 1967. The organization of genetic units in chromosomes. J. Cell Sci. 2:1–7.

Glassman, E. 1965. Genetic regulation of xanthine dehydrogenase in *Drosophila melanogaster*. Federation Proc. 24:1243–1251.

Hartman, P. E. and S. R. Suskind. 1965. Gene action. Prentice-Hall, Inc., Englewood Cliffs, N. J. 260 pp.

Huang, R. C. and J. Bonner. 1962. Histone, a suppressor of chromosomal RNA synthesis. Proc. Nat. Acad. Sci. U.S. 48:1216–1222.

Jacob, F. and J. Monod. 1961. Genetic regulatory mechanisms in the synthesis of proteins. J. Mol. Biol. 3:318–356.

Johri, M. M. and J. E. Varner. 1968. Enhancement of RNA synthesis in isolated pea nuclei by gibberellic acid. Proc. Nat. Acad. Sci. U.S. 59:269–276.

Lyon, Mary F. 1968. Chromosomal and subchromosomal inactivation. *In* Ann. Rev. Genet. 2:31–52. Annual Reviews, Inc., Palo Alto, Calif.

Mirsky, A. E. and H. Ris. 1951. The DNA content of animal cells and its evolutionary significance. J. Gen. Physiol. 34:451–462.

Ohno, S., U. Wolf and N. B. Atkin. 1968. Evolution from fish to mammals by gene duplication. Hereditas 59:169–187.

Schimke, R. T. 1967. Protein turnover and the regulation of enzyme levels in rat liver. Nat. Cancer Inst. Monograph 27:301–314.

Schimke, R. T., E. W. Sweeney and C. M. Berlin. 1965. The roles of synthesis and degradation in the control of rat liver tryptophan pyrrolase. J. Bio. Chem. 240:322–331.

Toft, D. and J. Gorski. 1966. A receptor molecule for estrogens: isolation from the rat uterus and preliminary characterization. Proc. Nat. Acad. Sci. U.S. 55:1574–1581.

Tomkins, G. M. and B. N. Ames. 1967. The operon concept in bacteria and in higher organisms. Nat. Cancer Inst. Monograph 27:221–234.

# SIX

## Chromosomal

## Differentiation

Chromosomes are among the most complex organelles found in metazoan cells. They have been recognized and studied by cytologists for more than a century, and their role as carriers of genetic information has been known since 1900. We now know that DNA is the essential chemical component for the storage of genetic information in the chromosome, but the DNA makes up only about half the total material in a chromosome. The remainder, mostly protein but also including a variety of other compounds, functions in unknown ways but probably regulates the readout of information from the DNA— that is, the transcription of RNA from DNA templates. Obviously, an understanding of the structure and function of chromosomes is central to the entire field of developmental genetics. Yet relatively little is known. Chromosomology remains one of the most challenging and potentially most fruitful areas of investigation for the present and near future.

We have already presented one way of studying chromosomes in our discussion of chromatin. We described the biochemistry of chromatin, obtained from interphase chromosomes, as studied by modern biochemical techniques. However, both electron microscopy and histochemistry have also contributed to our understanding of chromosome structure and function.

# The chemical composition of
# chromosomes

Most chemical analyses of chromosome composition require substantial amounts of nuclei as starting material. Such analyses have demonstrated that histones constitute a major component of chromosomes and have suggested that histones may be involved in genetic regulation. However, to examine the role of histones in the specificity of gene function requires methods sensitive enough to recognize histones in single nuclei. Histochemical staining reactions have been used for such refined analyses. Two stains, fast green and eosin, have proved particularly useful. Nuclei in a variety of tissues do not stain equally with the two reagents. In fact, in a given tissue a certain proportion of nuclei may be stained pink (eosin), others green, and a third category, in various shades of violet. The relative proportion of these two different nuclear staining reactions in a given tissue is constant and reproducible, indicating significant differences in chromosome structure among the cells. Even within a single nucleus, different chromosomes may stain differently. In spermatocytes of grasshoppers, for example, the bulk of the chromosomes stain green but the compacted chromosome stains purple.

The biological significance of these chemical differences—that is, what the stains demonstrate—is not well understood. But there is good evidence to indicate that eosin binds preferentially to the so-called lysine-rich histones, whereas fast green binds to the arginine-rich histones. During erythroid development, when RNA synthesis decreases, the nuclei become increasingly eosinophilic, indicating an increase in the lysine-rich histones on the relatively inert chromosomes. The histochemical approach is also useful for the study of the chemical composition of chromosomes in different functional states. Ideal material for such studies are *Drosophila* giant chromosomes, which exhibit some genes in action and others in an inactive state. The active genes transcribing RNA are often identified by large chromosomal puffs, whereas inactive chromosomal regions are characteristically compacted and do not synthesize RNA. According to the usual models of histone action, one might expect active loci to be devoid of histones, but this is not so. When puffed and unpuffed regions of giant chromosomes are stained for histones, the histone-to-DNA ratio in both is the same. These values are obtained by microdensitometric methods that are quite reliable. When puffed and unpuffed regions of a chromosome are exposed to an unspecific protein stain, which is not selective for histones, then the puffed regions exhibit a considerably higher ratio of protein to DNA. This is not the case after histone staining, and one is forced to conclude that the extra protein in puffed regions is of a nonhistone nature.

Chemical analysis of chromosomes reveals that the ratio of histone to nonhistone protein varies greatly in different cell types and in the same cell line at successive stages in cell differentiation. The histone content remains relatively constant but the content of nonhistone protein changes. Thus, one is led to the conclusion that the nonhistone protein is closely related to the differentiation of the chromosome, and therefore to gene function.

## Stability of differentiated state
## of chromosomes

Differential gene function as the central theme of cell differentiation presupposes that genes are turned off and on in accord with the physiological requirements of the cell. At the chromosomal level, this means that activated genes along a particular chromosome transcribe RNA whereas other, even adjacent, genes remain inactive. The giant chromosomes of dipteran larvae are ideal material to demonstrate directly that such differential gene function actually occurs. These giant chromosomes are composed of large numbers of parallel fibers, in some chromosomes over 1000, each fiber representing a chromatid. These chromosomes develop by progressive polytenization, and are often visible to the naked eye. They measure about 1 millimeter in length and some 50 microns in diameter. Because of the fidelity in pairing of homologous genetic sites, the sequences of genes along these fibers are in perfect register. This arrangement is reflected in a pattern of transverse bands, as shown in Fig. 6-1. Giant chromosomes, especially of dipteran salivary glands, were seen by cytologists of the last century. The banding pattern visible along these giant chromosomes was eventually accepted as cytogenetic confirmation of the genetic linkage data showing that genes are arranged linearly along the chromosomes. This conclusion was further borne out by experimental evidence demonstrating that X-ray induced chromosome deletions, visible at the cytological level, lead to phenotypic abnormalities similar to those produced by mutations that map to the same chromosomal region.

The cytologists of the last century reported early that the banding pattern of such giant chromosomes was very similar in the various

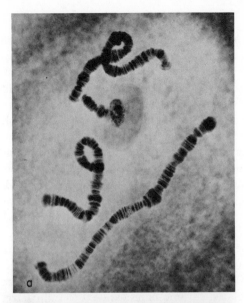

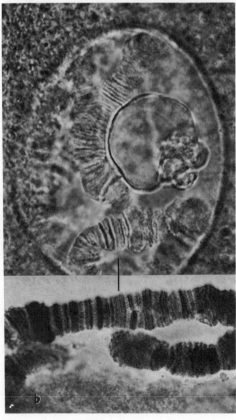

Fig. 6-1 (a) Giant salivary chromosomes of the gnat, *Chironomous,* showing the characteristic banding pattern.
Each chromosome is polytenic and may
consist of as many as 1,000 double
helical strands of DNA held side by side
in register by an as yet unidentified
mechanism. (b) The visible individual
bands on these chromosomes can be
related to identified genes. In this figure
a single band deficiency is shown in a
living chromosome at a magnification of
700 (above) and after fixation and
staining (below). One of the bands is
present in only ½ the chromosome. This
implies that a small deficiency was present in one of the homologous chromosomes that fused to form the giant
chromosome. Such deficiencies are
evident because of the precise lateral
association of these chromosomes. (Courtesy of D. F. Poulson.)

tissues of an organism. Geneticists later used this observation as a strong argument in favor of the genetic identity of different cell types. A few careful observers noticed, however, that sometimes it was difficult to identify certain sequences of bands; however, these difficulties were usually attributed to technical limitations and, at first, disregarded. These irregularities in banding patterns were consistently observed in some cell types but not in others, or at certain developmental stages but not at others. One type of irregularity, called a "puff," had the appearance of a loosened chromosomal structure. These puffs may be restricted to one single chromosomal band, or may include several adjacent bands. A very attractive hypothesis was soon formulated, which maintained that puffs reflected genetically active loci. And, indeed, when RNA precursor molecules labeled with the isotope tritium are injected into insect larvae, and chromosomes from various tissues processed for radioautography, the puffed regions of the chromosome (Fig. 6-2) are heavily labeled. More recently, RNA labeling has been observed even in isolated chromosomes exposed to radioactive precursor molecules, indicating that the chromosome itself contains the entire biosynthetic machinery required for RNA synthesis. A thorough survey of puffing patterns in various cell types throughout the development of *Drosophila* and also *Chironomus* shows that puffing patterns are, indeed, tissue and stage specific, in accord with the concept of differential gene function (Fig. 6-3).

In one case, the nature of the RNA synthesized at a particular puff

Fig. 6-2   RNA synthesis in chromosome puffs. The heavily labeled bands are more active in synthesizing RNA from the radioactive precursor. (Photograph courtesy of W. Beermann.)

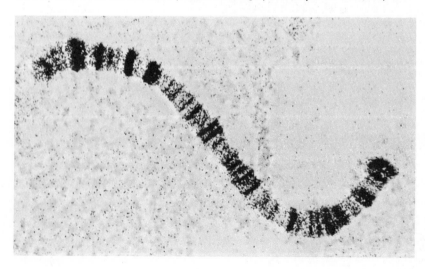

has been identified. By molecular hybridization techniques, it was shown that the nucleolar organizer region of *Drosophila* giant chromosomes produces ribosomal RNA. Through a series of selected crosses, *Drosophila* strains containing one, two, three, or four nucleolar organizer regions were produced. Then the amount of ribosomal DNA present in each genome was determined by measuring the amount of radioactive ribosomal RNA that could be hybridized to the respective DNA's. As Fig. 6-4 shows, the amount of ribosomal DNA is proportional to the number of nucleolar organizers.

In *Chironomus* was found a genetic locus responsible for the production of distinctive proteinaceous granules in salivary gland cells. Cytological examination of the salivary glands of two closely related species, *Chironomus tentans* and *Chironomus pallidivittatus,* showed that the glands of these two species differed in the appearance of a few cells. These cells, called "special cells," are located around the duct of the gland (Fig. 6-5). Fortunately, these two species can be hybridized successfully. When this genetic cross is made, the progeny contain fewer granules in the special cells than does the parent species *Chironomus pallidivittatus.* When the progeny are crossed inter se, the second generation consists of three classes of individuals in a ratio of 1:2:1— those with many granules in their special cells, those with fewer granules, and those with no granules at all. This indicates monofactorial Mendelian inheritance of the granules. The genetic locus responsible for this trait was determined by analysis of the giant chromosomes of the organisms in this cross. Figure 6-6 shows the results of this investigation. Notice that whenever the fourth chromosomes of an individual are both of the *C. tentans* type, the special cells lack the proteinaceous granules.

Detailed cytogenetic analysis indicated that the locus responsible for the production of the proteinaceous granules probably resided in a puff on the distal end of chromosome four. When chromosome four was carefully studied in both species, it was found that in *C. tentans,* no puff was present at that particular locus. In the first-generation hybrid between the two species, one puff was seen in that particular chromosomal region, and in *C. pallidivittatus,* two puffs were seen. This is the first case in which RNA synthesis by a known genetic locus has been correlated with the appearance of a chromosomal puff and with the appearance in the cytoplasm of a particular proteinaceous product. Although the chemical nature of this cytoplasmic protein is not yet known, the combination of cytogenetics, RNA autoradiography, and protein analysis should enable us to recognize the biological function of various messenger RNA's. In fact, in a recent analysis the protein controlled by a known locus on the chromosome has been

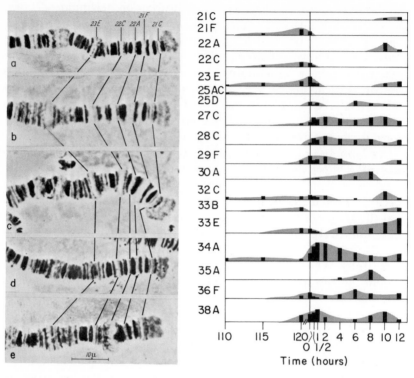

Fig. 6-3   Banding patterns of *Drosophila* salivary gland chromosomes. On the left (above) are shown actual photographs of identical chromosome regions at various developmental stages (a–e). On the right, cytogenetic regions (21c, 21f, etc.) are shown to be puffed (vertical bars) or unpuffed (no vertical bars) as a function of developmental stage in a diagrammatic way. The figure on the facing page presents diagrammatically the concept of differential gene function as applied to chromosome puffing patterns. (From M. Ashburner. 1967. Chromosoma 21:398–428. Fig. 1, 2.)

identified as a zone after acrylamide electrophoresis. Also it is already possible to localize on the genetic map of *Drosophila* the structural genes responsible for the production of enzymes. This is possible whenever mutant enzymes have altered electrophoretic mobility. Figure 6-7 provides an example in the form of the *Drosophila* enzyme, alcohol dehydrogenase. Notice that the enzyme appears in two electrophoretic patterns, labeled I and II, differing in their electrophoretic migration. The progeny of a cross between these two parents exhibits a different electrophoretic pattern of alcohol dehydrogenases, containing all the parental forms and, in addition, three hybrid molecules. Mendelian analysis locates the gene for this enzyme on the second chromosome; through cytogenetic techniques, the position of this locus on giant chromosomes has been narrowed to just a few bands. A number of

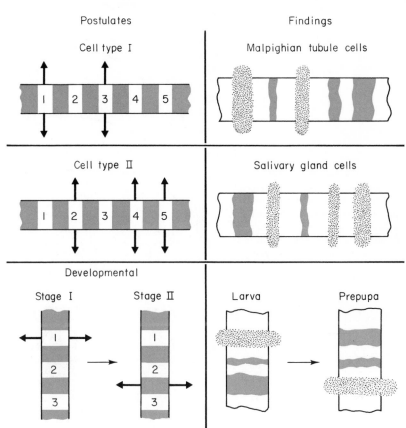

The Hypothesis of Differential Gene Activity

Postulates — Findings

Cell type I — Malpighian tubule cells

Cell type II — Salivary gland cells

Developmental

Stage I — Stage II — Larva — Prepupa

Fig. 6-3  Continued.

other loci responsible for the formation of enzymes with altered electrophoretic properties have also been found and are depicted in Fig. 6-7. For some of these enzymes, the tissue and stage specificity of appearance has already been elucidated, and it should be promising to investigate the synthesis of RNA at any of these genetic loci.

The enlarged, puffed regions of chromosomes do not, in general, contain more DNA than nonpuffed regions. An exception is the DNA puffs found in *Rhynchosciara* and *Sciara* chromosomes. These chromosomes were exposed to tritiated thymidine, a precursor of DNA, and then processed for radioautography. The results show clearly that certain regions of the chromosomes replicate their DNA more rapidly and more abundantly than do other regions. We mentioned earlier

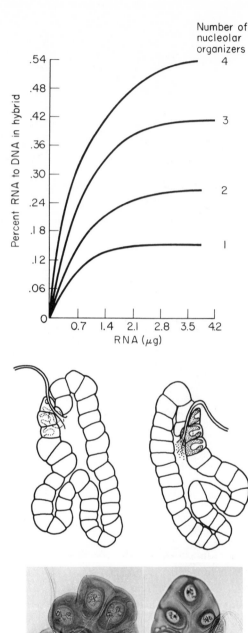

Fig. 6-4 The nucleolar organizer region of *Drosophila* chromosomes; site of ribosomal DNA. Amount of ribosomal RNA synthesized is proportional to the number of nucleolar organizers. (From Ritossa et al. Nat. Cancer Inst. Monograph **23**: 452.)

Number of nucleolar organizers

Drawing

Photograph

Fig. 6-5 The "special cells" of the salivary gland of *Chironomus*: first case of correlation of puff and gene product. Bottom, photograph. Top, drawing of section through gland showing granular gene product. (From W. Beermann. 1961. Chromosoma **12**:1–25. Fig. 1.)

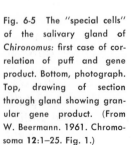

Chironomus tentans     Chironomus pallidivittatus

| Case | 2L | 2R | 3L | 3R | 4 | SZ |
|------|-----|-----|-----|-----|-----|-----|
| 1 | tt | tp | pp | pp | tp | + |
| 2 | tt | pp | pp | pp | pp | + |
| 3 | tp | tp | pp | pp | pp | + |
| 4 | pp | tp? | tt | tt | tp | + |
| 5 | tp | tp | tp | tp | pp | + |
| 6 | tt | tt | pp | pp | pp | + |
| 7 | tp | pp | pp | pp | tp | + |
| 8 | tp | pp | tp | tp | pp | + |
| 9 | tt | tp | pp | pp | pp | + |
| 10 | tp | tt | tp | tp | pp | + |
| 11 | tt | tp | tp | tp | tp | + |
| 12 | tt | tp | tp | tp | tt | − |
| 13 | tt | tt | pp | tp | tp | + |
| 14 | tt | tp | tt | tt | tp | + |
| 15 | tt | tp | pp | pp | tp | + |
| 16 | tp | tp | tt | tt | tp | + |
| 17 | tt | tp | tp | tp | tp | + |
| 18 | tt | tp | tp | tp | tt | − |
| 19 | tp | tt | tt | tt | tp | + |
| 20 | tp | tp | pp | pp | tp | + |
| 21 | tt | tp | tp | tp | tp | (+) |
| 22 | pp | pp | tt | tt | tp | + |
| 23 | pp | pp | tp | tp | tp | + |
| 24 | tp | tp | tp | tp | pp | + |
| 25 | tp | pp | tp | pp | tt | − |

Fig. 6-6 Genetic evidence for the location of the "special cell" trait. Genotype (t = tentans; p = pallidivittatus) of each chromosome arm (2L, 2R, 4) in 25 cases of $F_2$ of a cross between tentans and pallidivittatus SZ: special cells. (From W. Beermann. Table 1.)

that the nucleolar organizer of amphibians contains multiple copies of ribosomal DNA molecules. Perhaps the DNA puffs of insect chromosomes represent regions in which the genes have been replicated many times.

The control mechanisms underlying puffing are not understood. Since it was early recognized that puffing patterns differed from one larval stage to the next, it was hypothesized that molting hormones, such as ecdysone (see p. 169 for further discussion), might be regulating gene activity. Indeed, injection of ecdysone into larvae of *Chironomus* leads to the appearance of RNA puffs within less than an hour. The antibiotic actinomycin prevents this response of chromosomal loci to the injected hormone. This phenomenon has been

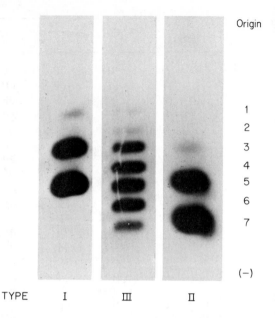

Fig. 6-7 (Top) Three different alcohol dehydrogenase phenotypes as revealed by electrophoretic resolution of tissue homogenates. Types I and II are homozygotes; type III is the heterozygote. (Bottom) Karyotype of *Drosophila* salivary chromosomes showing location of gene for alcohol dehydrogenase.

Origin

1
2
3
4
5
6
7

(−)

TYPE    I        III        II

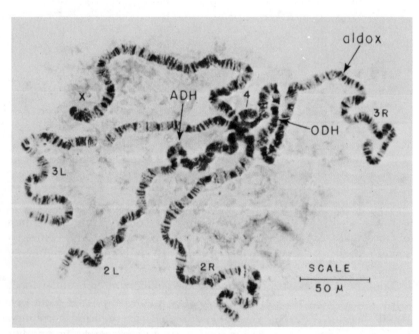

subjected to many different interpretations. Some investigators assume that ecdysone activates specific genes; others assume that ecdysone leads to a subtle change in the balance between sodium and potassium ions, which, in turn, is responsible for the activation of the respective loci.

Dipteran chromosomes are not the only giant chromosomes found in animal cells. Another type of large chromosome is the so-called lampbrush chromosome. This type of chromosome has been observed in the oocytes of numerous organisms. They have been seen in worms, mollusks, sea urchins, and insects. Among the vertebrates, they have been noted in cyclostomes, sharks, teleosts, urodeles and anurans, reptiles, and birds. Structures similar to oocyte lampbrush chromosomes have also been described during spermatogenesis in *Drosophila*. A typical lampbrush chromosome, as shown in Fig. 6-8, consists of a central longitudinal axis with a large number of beads or chromomeres from which lateral loops originate. Careful optical and enzymatic analyses indicate that the central axis consists of two double helices of DNA, that the beads or chromomeres represent areas along the chromosome in which supercoiling of the DNA occurs, and that the loops consist of one DNA double helix, with associated proteins and RNA. The pattern of loops along the chromosomal axis is specific, just like the banding pattern in a giant chromosome. Furthermore, each loop has a characteristic structure. The analogy between giant

Fig. 6-8    Lampbrush chromosomes. (Courtesy of J. G. Gall.)

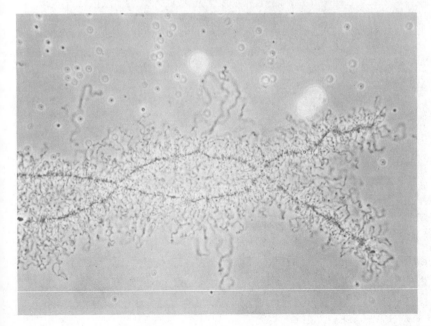

chromosomes and lampbrush chromosomes exists also at a functional level. When radioactive RNA precursors are administered to amphibians, for example, and their lampbrush chromosomes are then subjected to radioautography, the extent to which individual loops become labeled varies greatly, indicating loop-specific rates of RNA synthesis.

## The ultrastructure of chromosomes

The analysis of chromosome fine structure through conventional electron microscopy is difficult, largely for technical reasons. Electron micrographs of sections through nuclei or of whole-mounted

Fig. 6-9  Scanning electron micrograph of human metaphase chromosomes at (a) low and (b) high magnification.

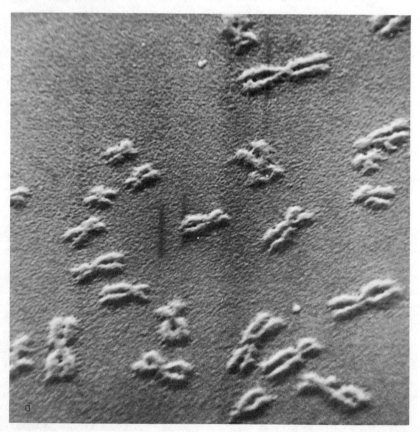

chromosomes reveal up to 64 fibrils that are largely oriented parallel to the longitudinal axis of the chromosome. Some workers have concluded that this arrangement reflects a varying degree of polynemy of chromosomes. Others maintain that the multiple fibers seen in electron micrographs are numerous folds of one and the same unit representing the longitudinal continuity of the chromosome. These different opinions agree in one respect, that is, on the dimension of the fibril. The fibril is a structure with a diameter of 100 Angstroms (A), often subdivided into two subunits, each of 40 A diameter. The weight of the evidence from observations with the electron microscope equates this 40-A fibril to a single DNA double helix with its associated protein. This simple arrangement makes excellent sense for a number of reasons. It explains the remarkable fidelity of replication and segregation of genetic information in meiosis and mitosis. It explains

Fig. 6-9  Continued.

furthermore the observed kinetics with which chromatids are digested by the enzyme deoxyribonuclease. In the case of amphibian lampbrush chromosomes, in which the latter experiments were performed, it would also agree with the length of a chromosome of known total DNA content. The haploid genome of salamanders, for example, is about 10 meters (m) long, which would be true if the chromatin indeed contains only one double helix of DNA per cross section.

A promising new approach for the study of chromosome fine structure is the use of the scanning electron microscope. In this instrument, electron reflection rather than electron transmission is used for the formation of images. Thus, the technique is particularly well suited for the study of surface configurations. Figure 6-9 shows a scanning electron micrograph of human metaphase chromosomes at low and at higher magnification.

## REFERENCES

Ashburner, M. 1967. Patterns of puffing activity in the salivary gland chromosomes of *Drosophila*. Chromosoma 21:398–428.

Beermann, W. 1964. Control of differentiation at the chromosomal level. J. Exp. Zool. 157:49–61.

Bloch, David P. 1966. Histone differentiation and nuclear activity. Chromosoma 19:317–339.

Clever, U. 1968. Regulation of chromosome function. *In* Ann. Rev. Genet. 2:11–30. Annual Reviews, Inc., Palo Alto, Calif.

Grossbach, U. 1969. Chromosomen-Antivität und biochemische Zelldifferenzierung in den speicheldrüsen von Camptochironomus. Chromosoma 28:136–187.

Pavan, C. 1964. Chromosomal differentiation. *In* Genes and chromosomes, Nat. Cancer Inst. Monograph 18:309–323.

Ritossa, F. M. and S. Spiegelman. 1965. Localization of DNA complementary to ribosomal RNA in the nucleolus organizer region of *Drosophila melanogaster*. Proc. Nat. Acad. Sci. U.S. 53:737–745.

Swift, H. 1964. The histones of polytene chromosomes. *In* J. Bonner and P. Ts'o [eds.] The nucleohistones, pp. 169–181. Holden-Day, San Francisco.

# SEVEN

## Translational and Epigenetic Control Mechanisms

### The longevity of messenger RNA

The rate of gene function—that is, the rate at which RNA is transcribed—is one of the limiting and controlling steps in the metabolic activity of a cell. The mRNA, once synthesized, might also function a variable number of times before being degraded, and the degree to which the mRNA molecule continues to function is obviously just as important in the synthetic activity of the cell as is the rate at which the RNA was initially synthesized. Moreover, the mRNA might be stored in an inactive form and accumulated over a period of time to be released for bursts of activity. In fact, considerable evidence has been produced to show that different species of mRNA molecules have very different half-lives, and also that many species of mRNA may be stored for long periods in the oocyte to be released to function after fertilization of the mature egg. The nature of the mechanisms utilized by the cell to prevent enzymatic degradation of mRNA is a very controversial subject indeed, and no resolution of the controversy can be expected until more conclusive evidence becomes available.

On the other hand, it is already clear that translation of RNA into protein in many cases is separated for lengthy periods from the initial transcription. This is true not only in the embryonic systems of the sea urchin and the frog, but also in differentiating cells. We choose the erythrocyte as an example.

## Erythroid development

The nomenclature of the morphological stages in erythrocyte development varies with different investigators, but these variations are irrelevant for our discussion. We are adopting one that is often used: stem cell→proerythroblast→basophilic erythroblast→polychromatophilic erythroblast→orthochromatic erythroblast→reticulocyte→erythrocyte (Fig. 7-1). The stem cells produce impressive numbers of erythrocytes. In man, for example, it has been estimated that every day $2 \times 10^{11}$ red blood cells are formed, each of which has a lifetime of about 100 days, leading to an average total red cell population of about $25 \times 10^{12}$ cells. The stem cells contain no hemoglobin. In subsequent stages of development, about two to three additional cell multiplications occur. Synthesis of RNA is observed principally until the basophilic erythroblast stage but not during later stages. Reticulocytes in particular do not synthesize RNA, which has been verified in several organisms by biochemical and autoradiographic techniques. But it is at the reticulocyte stage that protein synthesis, particularly hemoglobin synthesis, becomes very active. Therefore, the messenger RNA responsible for the synthesis of hemoglobin must have been produced long before its actual utilization for protein synthesis, and meanwhile been protected from enzymatic degradation.

The degree of stability of the hemoglobin RNA templates has been

**Fig. 7-1** The developmental history of red blood cells. A population of multiplying stem cells gives rise to more stem cells (S) and differentiating stem cells (S′), which through a series of transformations (E′–E′′′), ultimately give rise to reticulocytes and red blood cells. (From E. Goldwasser. 1966. *In* Current topics in developmental biology 1:173–210. Fig. 4.)

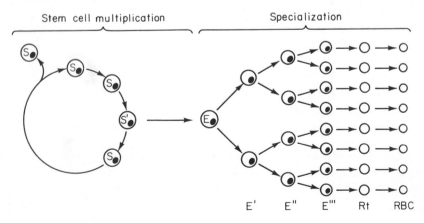

Stem cell multiplication          Specialization

E′     E″     E‴     Rt     RBC

determined experimentally. Labeled RNA precursors were administered to rabbits, and autoradiographs were made of blood smears prepared at various times after the pulse. Almost 2 days are required for the appearance in the blood of reticulocytes containing radioactive RNA. We know that the circulating reticulocyte does not synthesize RNA; in fact, in mammals it no longer contains a cell nucleus. Consequently, its radioactive RNA must have been produced at a previous stage of maturation. We may conclude that the template RNA is synthesized during or prior to the basophilic erythroblast stage and survives for at least a 2-day period.

Since during erythroid development transcription and translation are well separated in time, this cell system should be a good place to search for a mechanism of messenger RNA stabilization. Many investigators have done this, and interest has been focused on ribosomes. When different stages of erythroid development were examined with the electron microscope, it was found that the relative amounts of single ribosomes and polyribosomes differed according to the cell stage. In nucleated stages, only 1% of the ribosomes occur as single ribosomes, 43% being in the form of polysomes containing four ribosomes. In early reticulocytes, on the other hand, this ratio is closer to 1:1, and in later stages single ribosomes again predominate. The concentration of ribosomes declines sharply with the maturation of the reticulocytes. The finished erythrocyte contains no ribosomes at all. Thus, the messenger RNA synthesized in erythroblasts becomes associated with ribosomes to form polysomes, and it remains stable in this form until it is utilized for protein synthesis in the early reticulocyte stage of erythrocyte development. This mechanism of messenger RNA stabilization has also been postulated for other cell systems—for example, in the differentiation of the lens in the eyes of vertebrates.

### Lens differentiation

The lens of the vertebrate eye is an aggregate of fibers, each formed during terminal differentiation of an epithelial cell. The lens fiber cell—like the mature erythrocyte—lacks a nucleus. The biochemical events accompanying fiber differentiation have been worked out in considerable detail at both nucleic acid and protein levels of metabolic activity. For the purpose of this discussion, we shall simply describe the experiments that demonstrate the stabilization of messenger RNA in lens cells. These experiments were carried out on calf eyes, the lenses of which contain both differentiated fibers and epithelial cells. The two cell types were isolated from one another and incubated in media containing either radioactive amino acids or radioactive RNA precursors. The protein resulting from this in vitro synthesis was then analyzed by chromatography. This method resolves three protein species, referred to as alpha, beta, and gamma crystallins. Notice in Fig. 7-2 that epithelial cells synthesize all three species, provided that concomitant RNA synthesis is not prevented by actinomycin. If this inhibitor is present, synthesis is greatly reduced. If the same pair of experiments is carried out using fiber cells, synthesis of the three crystallins occurs even in the presence of the drug. If in such experiments RNA, rather than protein, synthesis is measured, then it is found that RNA is synthesized not in fiber cells but in epithelial cells in the absence of actinomycin, but not in the presence of this inhibitor. These results strongly indicate that the templates necessary for protein synthesis in lens differentiation are synthesized early and stabilized during differentiation.

Rather than expanding this list of examples of stabilization of templates, let us now try to assess the amount of time that RNA may be stabilized. In the classical view, mRNA was regarded as a molecule that started to decay with first-order kinetics as soon as it was made. Such short-lived messenger RNA is characteristic of bacteria and has been observed in animal cells also. Using a pulse label in the presence or absence of actinomycin, the half-life of the messenger RNA for amino levulinic acid synthetase in rat liver has been estimated to be of the order of 1 hr. At the other extreme lie the cases of template RNA molecules found in dormant plant seeds. No protein synthesis occurs in the dormant seed. But when seeds are placed in water, polysomes are formed within just a few hours, even in the absence of RNA synthesis, and protein synthesis is initiated. One is led to conclude that the seed contains stable templates. Since many seeds may be kept dormant for years, the life span of template RNA contained within them must be very long indeed. In animal cells, extremely long life spans have been noted for RNA templates in sea urchin and frog oocytes. Between the two extremes lie the cases that might be termed "moderately long-lived" messenger RNA, such as RNA's responsible for hemoglobin and lens fiber synthesis.

It is doubtful that a fixed messenger RNA half-life is characteristic of a particular cell line. In fact, there is evidence in the literature that a given cell type—for example, mammalian liver cells—contains messenger RNA molecules of widely different degrees of stability. It is

Fig. 7-2 (a) The phases of cell differentiation in lens development. (b) Crystallin synthesis in lens epithelial cells in the absence and presence of actinomycin-D. (c) Crystallin synthesis in lens fiber cells in the absence and presence of actinomycin-D. (From J. Papaconstantinou. 1967. Science 156:338–346. Figs. 1, 8, 9, 10, 11. Copyright 1967 by the American Association for the Advancement of Science.)

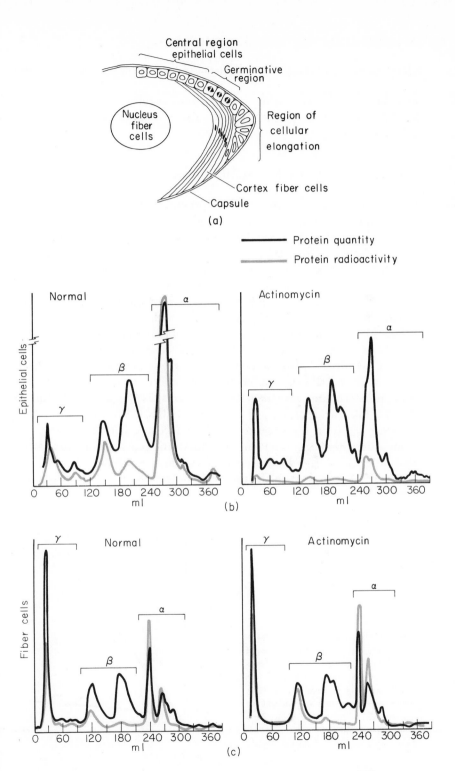

(a)

Central region
epithelial cells

Germinative region

Nucleus fiber cells

Region of cellular elongation

Cortex fiber cells

Capsule

———— Protein quantity
———— Protein radioactivity

Normal

Epithelial cells

γ    β    α

Actinomycin

γ    β    α

(b)

Fiber cells

Normal

γ    β    α

Actinomycin

γ    β    α

(c)

115

clear, however, that separation in time of transcription and translation is a regulatory mechanism quite widely used by higher organisms. To elucidate the molecular mechanisms responsible for the differential longevities of messenger RNA molecules is another challenge of contemporary biology.

## Mechanisms of mRNA stabilization

In Ch. 3, we described the protein-synthetic pattern of early sea urchin development. A glance at Fig. 3-1 will refresh your memory that the rate of protein synthesis increases sharply after fertilization, and, furthermore, that this abrupt increase occurs even in the absence of RNA synthesis. The rapid metabolic response of eggs after fertilization led investigators to assume early that postfertilization protein synthesis was achieved on RNA templates previously stored in the egg during oogenesis. Direct evidence is now available that template RNA is indeed present, although quiescent, in the unfertilized egg.

RNA extracted from unfertilized eggs shows template activity for protein synthesis when added to cell-free extracts prepared from bacteria. Such cell-free extracts have the ability to incorporate amino acids into proteins, provided that template RNA is added. Notice in Fig. 7-3 that incorporation of radioactively labeled serine into protein does not occur in the absence of added RNA. When RNA prepared from unfertilized eggs is added, incorporation occurs readily, indicating that unfertilized eggs do, indeed, contain template RNA.

If mRNA is present in the unfertilized egg, why is it not utilized for

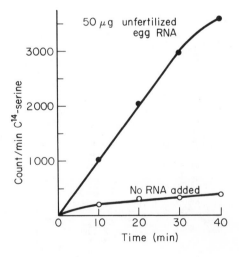

Fig. 7-3 Template activity of unfertilized sea urchin eggs. RNA prepared from unfertilized eggs causes radioactive serine to be incorporated into protein in a cell-free protein synthesizing system, indicating that the unfertilized egg contains informational RNA. (From D. W. Slatter and S. Spiegelman. 1966. Proc. Nat. Acad. Sci. U.S. 56:164. Fig. 4.)

protein synthesis? Obviously, translation of this mRNA is prevented in one way or another. Investigators in numerous laboratories have therefore begun to analyze unfertilized eggs for the presence or absence of components of the translational apparatus.

The general conclusion from these investigations has been that the so-called postribosomal fraction of unfertilized sea urchin eggs—that is, the cellular components that do not sediment as quickly as ribosomes during centrifugation—are capable of promoting protein synthesis. The defect in protein synthesis thus appears to be localized in ribosomes or in the heavier fractions of the cell. This is evident from an experimental result portrayed in Fig. 7-4. In this experiment, microsomes—that is, particles containing ribosomes and other particles, obtained either from unfertilized eggs or from two-cell stages—were incubated with a postribosomal fraction that itself was obtained from one or the other of those developmental stages. Radioactive leucine was present during incubation, and its incorporation into protein was measured. Appreciable incorporation did not occur when the microsomes were obtained from unfertilized eggs, regardless of whether fertilized or unfertilized postribosomal fractions were present in the incubation mixture. Also, appreciable incorporation of leucine *did* occur when ribosomes were obtained from two-cell embryos, again regardless of the source of the postribosomal fraction. From this evidence, the block to protein synthesis seems to reside in the microsomal fraction.

Attention has since been focused on this fraction. It contains both

Fig. 7-4 Incorporation of leucine into protein in an in vitro protein synthesizing system involving different cell fractions.

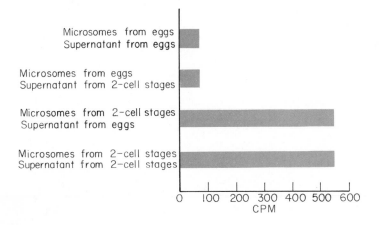

ribosomes and template RNA, but in a form that cannot be translated into protein in the unfertilized egg. When such a microsomal fraction is pretreated with trypsin and then used in a cell-free incubation mixture, like the one shown in Fig. 7-4, incorporation of amino acids into protein takes place. This result has led to the conclusion that ribosomes in unfertilized eggs are blocked by a trypsin-sensitive inhibitor. On fertilization or activation of the egg, the mRNA and the translational machinery apparently become released to function actively. It remains a challenge to discover just how the templates in the egg are stabilized in the first place, and how egg activation leads to their release. Perhaps the best way to approach this question is by presenting a hypothetical arrangement for the storage of mRNA in the cell (Fig. 7-5) and then proceed to describe the evidence relevant to this scheme.

According to this scheme, mRNA produced in the nucleus is stored in cytoplasmic informosomes. Informosomes are particles containing RNA and protein; they may be isolated by prolonged centrifugation of postribosomal supernatants. They move more slowly in the centrifugal field than do ribosomes, but faster than protein and RNA molecules. Typically, informosomes fall into particle size classes ranging from 20 S to 65 S. Six or seven peaks of such particles have been obtained in preparative centrifugation of cell homogenates from cleavage stages of the sea urchin and the loach, *Misgurnus fossilis*. These particles have template activity when introduced into a cell-free protein-synthesizing

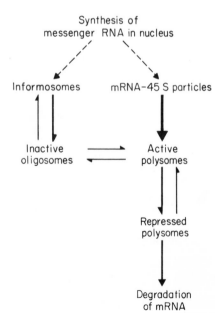

Fig. 7-5 Hypothetical scheme of mRNA stabilization.

system, indicating that they contain informational or messenger RNA. Apparently each of the six or seven particle classes contains protein and a particular size class of RNA. The 31 S particle, for example, was found to contain 13 S RNA; on the other hand, the 49 S particle contained 22 S RNA. By the use of molecular hybridization techniques, it was found that the RNA contained in these particles readily hybridized with sea urchin DNA, which again indicates that the RNA of the particles is informational in nature—hence, the name informosomes.

The presence of informosomes was first established in total homogenates of embryos that had been incubated in media containing radioactive RNA precursors. The various size classes of informosomes all contained newly synthesized, but stable, RNA.

Another component of these aggregates is the so-called oligosomes. They have sedimentation constants of 90 S to 200 S and may form or be derived from informosomes. In pulse experiments, with labeled precursors of RNA, newly formed RNA becomes associated with these oligosomes rather than with the heavier fraction, that is, the polyribosomes. Polyribosomes sediment at 200 S or even faster, and it is on these particles that protein synthesis occurs, as witnessed by the incorporation of radioactive amino acids during protein synthesis. It should be noted that the heavy polyribosomes do not become readily associated with newly formed mRNA but are apparently programmed by preexisting messages. This observation led to the speculation that new mRNA is stored temporarily in inactive complexes with oligoribosomes. Specifically, it was postulated that informosomes form complexes with ribosomes, thus giving rise to the oligoribosome fraction (Fig. 7-5). If the informosomal protein coat that covers these complexes were removed, active mRNA-ribosome particles would be freed to form polysomes, and consequently to initiate protein synthesis. According to this model, it would be the inactive oligosomes (ribosomes plus informosomes) that are derepressed by the removal of the protein coat, as previously discussed. The resulting particles, devoid of their protein coat, would form polysomes that are active in protein synthesis. This concept is consistent with the observation that the number of active polysomes increases after fertilization and throughout the cleavage stages.

## Epigenetic modification of proteins

The regulation of protein structure and function does not stop with the translation of mRNA into a polypeptide chain. At that time, epigenetic modification of protein structure commonly occurs. Since the protein molecule is very large, it may be folded in different ways to generate a variety of three-dimensional conformational states, each with its own characteristic properties. Many physical chemists question the likelihood of a protein molecule existing in alternate stable conformations. Nevertheless, increasing evidence points to the existence of enzymes in such alternate states. However, these different conformations might stem from the attachment of ligands to the protein to alter its conformation.

Enzymes are very commonly polymers, and one type of control is the regulation of polymer size. For example, the enzyme glutamate dehydrogenase exists as a polymer series with molecular weights that vary from 250,000 to 1 million. The state of aggregation of this particular enzyme is influenced by the concentration of the enzyme itself and by a variety of small molecules, such as steroids, nucleotides, and coenzymes. All these substances affect polymer size in vitro, and probably the same kinds of molecules regulate the state of aggregation of the enzyme within the cell. The state of aggregation is important in regulating the function of this enzyme in the metabolic pathways of which it is a part.

One type of epigenetic modification involves the deletion of a part of the peptide chain. This is clearly so for such enzymes as trypsinogen and procarboxypeptidase. Both of these proenzymes are activated by proteolytic removal of an inhibiting peptide fragment, clearly an epigenetic modification. Changes in the state of oxidation or reduction also affect the structure of a variety of enzymes. Malate dehydrogenase and lactate dehydrogenase are both affected by such reagents. These enzymes exist in a number of conformational isozymic forms, which can be interconverted through the use of oxidizing and reducing agents (Fig. 7-6). Epigenetic modifications sometimes involve the attachment of a small molecular moiety to the enzyme itself, as clearly shown by the sialic acid residues attached to alkaline phosphatase. The sialic acid generates a variety of electrophoretically distinct molecular forms of the enzyme, each form presumably fulfilling a somewhat different role in cell metabolism. The sialic acid may be removed from the enzyme by neuraminidase treatment, and all forms of the enzyme are then converted to a single more slowly migrating form. Thus, the initial differences in mobility can be attributed to different amounts of bound sialic acid. The same situation appears to hold for acid phosphatase extracted from various human organs. These also can all be reduced to the same electrophoretic variety by treatment with neuraminidase.

Glycogen phosphorylase likewise exists in a variety of isozymic forms, in this case generated by the degree of phosphorylation of the enzyme itself. Several isozymic forms of this enzyme are generated by complex

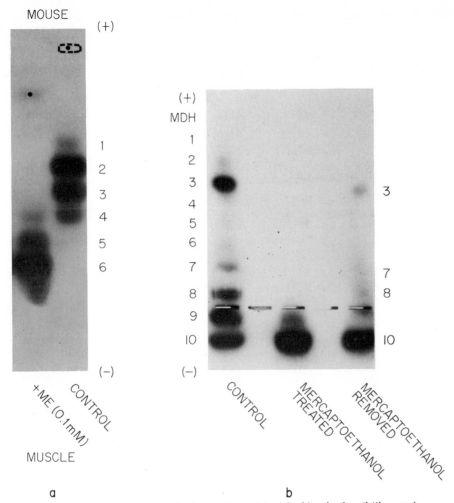

Fig. 7-6 Zymograms showing subbands of mouse LDH-5 (a), and subbands of snail (*Ilyanassa*) malate dehydrogenase (b). Note the redistribution of enzyme activity into different bands after reduction by mercaptoethanol. The effect is reversible and demonstrates that these subbands are epigenetic modifications of a single protein, both interconverted by mercaptoethanol treatment. [(b) from Archives of biochemistry. 1967. Vol. 122, p. 763. Fig. 8.]

interactions involving polymerization and combination with phosphate groups. These various isozymic forms of glycogen phosphorylase do not all have the same specific activity, and the interconversion of one to another regulates the total phosphorylase activity within the cell, Thus, a variety of epigenetic transformations of single polypeptide chains have been identified. These permit an amplification of the number of isozymic forms of an enzyme, each with characteristic kinetic properties and probably with a distinctive role to play in cell metabolism.

REFERENCES

Goldwasser, E. 1966. Biochemical control of erythroid cell development. *In* A. A. Moscona and A. Monroy [eds.] Current topics in developmental biology, vol. I, pp. 172–212. Academic Press, New York.

Papaconstantinou, J. 1967. Molecular aspects of lens cell differentiation. Science 156:338–346.

Slater, D. W. and S. Spiegelman. 1966. An estimation of genetic messages in the unfertilized echinoid egg. Proc. Nat. Acad. Sci. U.S. 56:164–170.

Spirin, A. S. 1966. "Masked" forms of messenger RNA in early embryogenesis and in other differentiating systems. *In* A. A. Moscona and A. Monroy [eds.] Current topics in developmental biology, vol. I, pp. 1–38. Academic Press, New York.

Tyler, A. 1967. Masked messenger RNA and cytoplasmic DNA in relation to protein synthesis and processes of fertilization and determination in embryonic development. *In* M. Locke [ed.] Control mechanisms in developmental processes, pp. 170–226. Academic Press, New York.

# EIGHT

## Nucleocytoplasmic

## Interactions

### The driving force of cell differentiation

Cell differentiation is based upon the differential activation of genes. Although the molecular mechanisms involved are not known, certain general relationships seem to be necessary. If the environment of the genes in the egg were completely uniform, it would be very difficult to imagine a mechanism that could generate differences among the blastomeres of the early embryo. Yet in many embryos these blastomeres have been demonstrated experimentally to be different from one another. Some destabilizing force must be introduced into the cleaving egg to set in motion the evident train of sequential changes. All such change must stem from the structure of the fertilized egg. Of course, eggs are not homogeneous cells. The cytoplasm is an exceedingly heterogeneous mixture of substances organized into many specific structures. Moreover, the distribution of these materials in different parts of the egg cytoplasm is neither uniform nor random (Fig. 8-1). Commonly, the vegetal hemisphere of the egg is more heavily laden with yolky materials than the animal hemisphere. Other nonrandom distributions of materials are evident in pigment and viscosity gradients and in the visibly different distributions of cell organelles.

Thus, the genes in the nuclei of the early blastomeres of a developing embryo necessarily find themselves in diverse cytoplasmic environments. Presumably, these different environments result in the activation of different sets of genes. The active genes then generate new cytoplasmic environments, which react with the genome of the

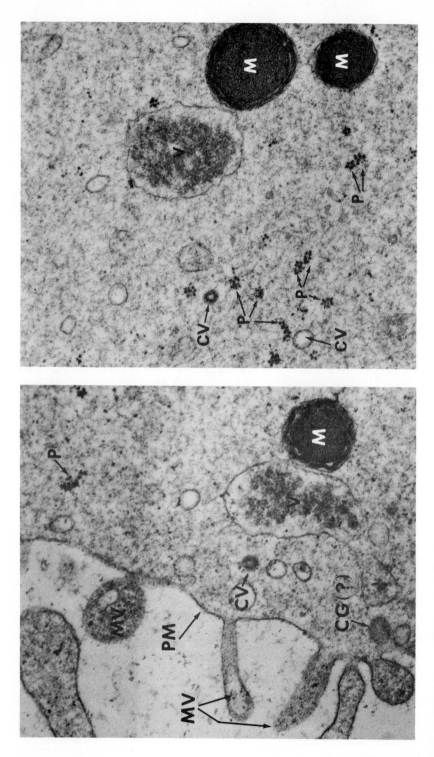

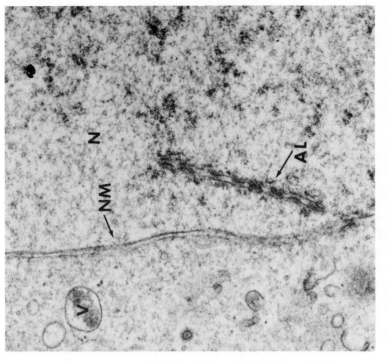

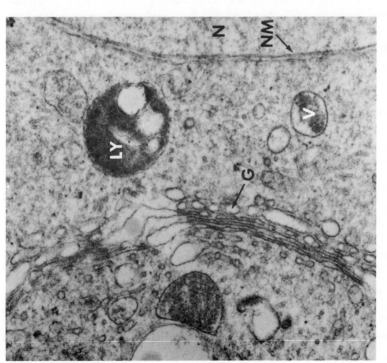

Fig. 8-1 The four electronmicrographs in this figure were selected to demonstrate representative areas in a fertilized rabbit egg. The panels proceed (left to right) from the outside of the egg to the nucleus. The contents of the egg are obviously exceedingly heterogeneous. The few structures that are labeled represent only a trivial proportion of the identifiable structures of a mammalian egg. CG, Cortical Granule; M, Mitochondrion; AL, Annulate Lamella; P, Polysome; V, Cytoplasmic Vesicles; MV, Microvillus; N, Nucleus; CV, Coated Vesicles; G, Golgi; LY, Lysosome; NM, Nuclear Membrane; PM, Plasma Membrane. (Courtesy of Mildred Gordon.)

125

cell to bring about the selective activation or inhibition of new groups of genes. Hence, a sequence of reciprocal reactions is inaugurated between the genome and the changing cytoplasmic environment so that the cell is driven along a predestined path of differentiation to the terminal cell type in the adult. It is, of course, possible to alter the fate of a cell by experimentally relocating the cell in a different tissue environment. These different tissue environments must induce changes in the cytoplasmic composition of the transplanted cell so that, again, different sets of genes are brought into play. This conceptual framework, within which one can understand the relationships between the genes and their chemical environment, is simple enough, but as yet we have essentially no knowledge of the specific molecular events involved. The identification of these molecular events constitutes one of the most challenging areas of current experimentation in developmental genetics.

## Molecular exchanges between nucleus and cytoplasm

Recent advances in molecular biology now make it clear that part of the exchange between the nucleus and cytoplasm is in the form of the various kinds of RNA molecules elaborated by transcription from nuclear DNA. At least a fraction of these RNA molecules emigrate into the cytoplasm and there participate in the formation of proteins that reflect the information contained in the activated genes. This part of the molecular traffic between nucleus and cytoplasm is relatively easy to recognize. The reverse flow from cytoplasm to nucleus is much more obscure. Yet, it must be rich in informational content. Molecules from the cytoplasm must in some way be responsible for the selective activation of nuclear genes. Only molecules as complicated as nucleic acid or protein seem to possess adequate informational content to fulfill this role. Thus, a shuttling of information-rich molecules back and forth between the nucleus and cytoplasm seems essential. In fact, recent experiments on *Amoeba proteus* have demonstrated just such a class of proteins. Amoebae offer unusual advantages because nuclei can easily be transplanted among them. The transplantation procedure involves the use of a microprobe to shove the nucleus from one amoeba into another directly through both cell membranes. In such microsurgery, the amoebae are placed adjacent to one another so that the transplanted nucleus is not exposed to the extracellular medium. Such transplanted nuclei continue to function normally. They and their molecular constituents can be identified by radioactive labeling (Fig. 8-2), achieved by feeding suitably labeled food material

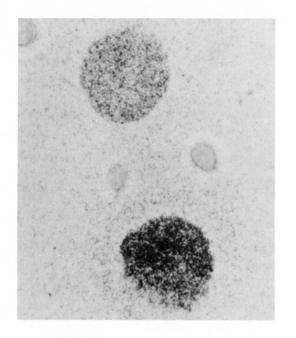

Nucleocytoplasmic protein
interactions

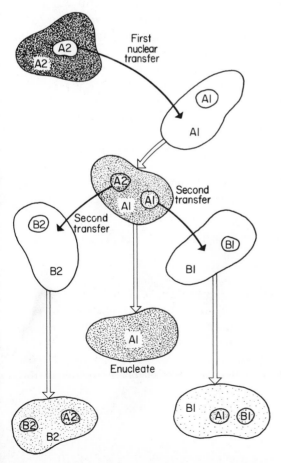

Fig. 8-2 Diagram showing Goldstein's nuclear transplantation experiment with labeled nuclei in Amoeba. (Top) Autoradiograph of part of an amoeba (out of focus) into which was grafted a lysine-$^{14}$C-labeled nucleus 20 hr before the cell was fixed. Radioactivity is almost completely localized over grafted nucleus and host cell nucleus. The most reasonable interpretation is that there exists a class of proteins (RMP) shuttling back and forth between the nucleus and the cytoplasm, but which is present in the former in a much higher concentration than in the latter. (Bottom) A nucleus (A2) from a $^3$H protein cell is grafted into an unlabeled cell (A1) for the first nuclear transfer. The next day, nucleus A2 is grafted into unlabeled cell B2 and nucleus A1 is grafted into unlabeled cell B1. Enucleate A1 is saved for assay. Either the two cells after the second transfer are immediately fixed whole to assay for total activity of nucleus A2 and of nucleus A1, or the four nuclei are isolated the next day to determine the distribution of $^3$H protein between nuclei B2, A2, A1, and B1. (Courtesy of Lester Goldstein.)

127

to the donor amoebae. The radioactivity in the nuclei of such cells can be assayed either by autoradiography (Fig. 8-2a) or by a Geiger counter after nuclear isolation. One dramatic observation from such experiments is that radioactive proteins from the transplanted nucleus soon find their way via the cytoplasm into the host nucleus (Fig. 8-2b). Moreover, the amount of radioactive protein in the two nuclei, host and transplanted, soon reaches a rather steady ratio. After equilibrium has been reached, the amount in each nucleus is approximately equal to the total radioactive protein that has remained in the cytoplasm. Since the nuclei have much smaller volumes than the cytoplasm, the concentration of radioactive proteins in the nucleus is far greater, about 50 times greater, in fact. The simplest interpretation of these observations is that some intact proteins leave a radioactive nucleus, migrate into the cytoplasm, but then return, either to the original grafted nucleus or to the host nucleus. The behavior of these proteins is what might reasonably be expected of a class of proteins responsible for conveying information from the cytoplasm to the nucleus.

In these experiments, not all the nuclear proteins were found to be so mobile. In fact, about 60% of the total labeled protein was relatively fixed in the nucleus. This protein did gradually leave the nucleus but at a much slower rate than the 40% that shuttled back and forth to the cytoplasm. These "slow turnover" proteins are probably composed of such classes as the proteins making up the nuclear envelope, structural proteins of the nucleolus, and various chromosomal proteins, such as the histones. However, even these slow turnover proteins do migrate into the cytoplasm, and some of them may return to the nucleus. In any event, it seems likely that the cytoplasm rather than the nucleus itself is the location for the synthesis of such proteins. Their behavior makes them unlikely candidates for mediating nuclear gene transcription. This role seems better assigned to the rapidly migrating proteins.

## Control of macromolecular synthesis in the nucleus by cytoplasm

In the preceding section, evidence was presented that proteins shuttle back and forth between the nucleus and cytoplasm and presumably serve as regulators of genetic transcription. None of these experiments proved that these proteins regulated genetic transcription. However, two classes of experiments have been devised that clearly demonstrate the dependence of nuclear function on the nature of the cytoplasm. These are based on experimentally formed combinations of nuclei and cytoplasm. The first involves nuclear implants in amphibian eggs. The second involves fusing somatic cells in vitro to make novel combinations of cytoplasm and nucleus at different stages in their respective developments. In addition to these experimental arrangements to show nuclear dependence on cytoplasm, several normally occurring events in cell differentiation in certain organisms likewise demonstrate a clear dependence of nuclear function on the cytoplasm. Such dependence is evident in the inactivation of one of the X chromosomes in female mammals (see p. 77).

The technique of nuclear transplantation in amphibian eggs will be discussed at greater length in a subsequent section. For the present discussion, it is sufficient to know that the nuclei of various kinds of cells, even of fully differentiated cells, may be transplanted into enucleated oocytes by microsurgical techniques. Nerve cells do not normally divide in the adult vertebrate, and, accordingly, the nuclei of these cells do not synthesize DNA. When nuclei are taken from adult frog brain cells (some of these may not be from nerve cells) and injected into oocytes with intact germinal vesicles, no DNA synthesis occurs in these injected nuclei, even when left in place for several days. In these experiments, tritiated thymidine was available within the oocyte; any DNA synthesis would have resulted in the incorporation of this precursor and would have been detected by autoradiography. When the brain cell nucleus was implanted in the oocyte at the time the germinal vesicle broke down, DNA synthesis occurred in this implanted nucleus. Apparently the germinal vesicle releases into the cytoplasm some factor responsible for initiating the synthesis of DNA in nuclei—even in nuclei that would normally never synthesize DNA. This factor may be DNA polymerase, a conclusion reached because it was possible to induce DNA synthesis in the cytoplasm of oocytes by injected purified DNA unassociated with any nucleus. Injection of the same materials preceding the breakdown of the germinal vesicle did not lead to any synthesis of DNA. These reactions are not specific because nuclei obtained from adult mouse liver cells behave similarly when injected into frog oocytes.

The interaction between the nucleus and cytoplasm that results in nuclear synthetic activity also involves physical changes in the nucleus. Implanted nuclei are relatively small compared with blastomere nuclei, but they swell rapidly in the mature oocyte. The sperm nucleus increases in volume more than 50 times within 30 minutes after fertilization, and various somatic nuclei implanted into the egg also increase greatly in volume, depending upon their size at implantation.

Even large blastula nuclei increase threefold. Smaller nuclei from the epithelial lining of *Xenopus* larval intestines increase 40 times, and adult brain nuclei increase 60 times. The cause of this swelling is unknown, but it appears to be an indispensable part of the interaction with the cytoplasm preceding the synthesis of nuclear DNA.

In addition to inducing DNA synthesis, the cytoplasm also controls the nature of RNA synthesis carried on by the nucleus. If nuclei actively synthesizing ribosomal RNA are transplanted into enucleated eggs, they cease this synthesis in accord with the fact that nuclei in early embryonic cells do not synthesize ribosomal RNA. The synthesis of ribosomal RNA is resumed later, when the embryo reaches the stage at which such synthesis is a normal part of development. The same general results were also obtained with reference to the synthesis of transfer RNA, which is likewise not synthesized by nuclei at very early stages in development. An interesting variant on these experiments was achieved with midblastula nuclei, which normally synthesize very little RNA although they are in the midst of rapid replications of DNA. Injection of these nuclei into immature oocytes with intact germinal vesicles resulted in the immediate synthesis of large amounts of RNA and the cessation of DNA synthesis. Thus, their metabolic behavior corresponded to the instructions known to exist in the cytoplasm of the oocyte.

The behavior of chromosomes in the mealy bug provides a striking example of extreme specificity in the interaction of chromosomes with the cytoplasmic environment. Both the paternal and maternal sets of chromosomes in the cells of female embryonic mealy bugs behave alike, as is generally true in nearly all cells. However, in male embryos, the paternal set of chromosomes is made heterochromatic and contributes very little to the genetic characteristics of the male (Fig. 8-3). The maternal set remains euchromatic and functional. Thus, in eggs destined to produce males, something in the cytoplasm must distinguish these two sets of chromosomes from one another and induce the physicochemical changes identified as heterochromatization (in the male set) and euchromatization (in the female set). In certain tissues of some species of male mealy bugs, the heterochromatic configuration of the chromosomes is reversed and both the maternal and paternal sets become euchromatic and functional. Thus, there is nothing intrinsically deficient in the paternal set of chromosomes, although this set must be recognizably different from the maternal set. Clearly a sensitive, highly specific molecular mechanism must exist in the cytoplasm of mealy bug eggs to distinguish homologous chromosomes and to restrict the heterosynthetic activity of the paternal set. The nature of the postulated molecular mechanisms involved is at present unknown but offers

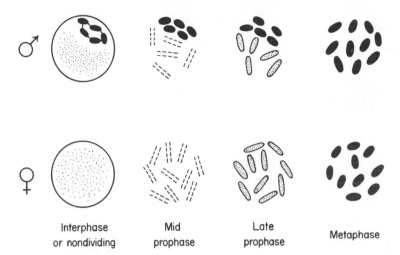

Interphase or nondividing     Mid prophase     Late prophase     Metaphase

Fig. 8-3 Diagram showing behavior of paternal set of chromosomes in the mealy bug. Mitosis in female and male mealy bugs. The female and the male mealy bug have the same number of chromosomes (ten is the commonest number). The chromosomes of the female (bottom row) remain unchanged during development. Half the chromosomes of the male (top row) become heterochromatized at an early embryonic stage and remain so throughout development. The heterochromatic chromosomes clump together to form a chromocenter at interphase and appear more condensed than the others during prophase. At metaphase all chromosomes of both sexes are equally condensed.

dramatic opportunities for investigating the specificity of interaction between chromosomes and their cytoplasmic environment.

Several lines of evidence lead to the conclusion that heterochromatin is relatively nonfunctional in contrast to euchromatin. Observations on genetic characteristics of male mealy bugs and on the expression of X-linked genes in female mammals clearly indicate that heterochromatic chromosomes are essentially nonfunctional. Heterochromatization is a reversible chromosomal condition, however, and is under the effective control of the cytoplasmic environment.

Nuclei or chromatin may be isolated and extracted from different kinds of differentiated cells, and their RNA transcribing ability tested. As might be expected, they exhibit the synthetic activities of the cell type from which they are prepared (Fig. 8-4a–c). Apparently, the selectivity of this chromatin-directed synthesis of RNA is caused by the protein moiety of chromatin. For, if chromatin is deproteinized experimentally and then tested for its transcriptional activity, it is seen to transcribe a much wider variety of RNA's than untreated chromatin does (Fig. 8-4d). Such deproteinized chromatin has the same synthetic activity as the corresponding DNA.

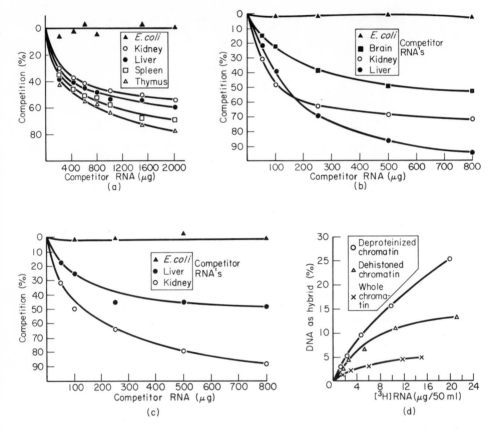

Fig. 8-4   Tissue specificity of RNA synthesis is remembered by isolated nuclei and by isolated chromatin. (a) RNA made by isolated thymus nuclei was hybridized to DNA in the presence of unlabeled competitor RNA prepared from other tissues. Notice that competitor RNA obtained from thymus is a more effective competitor than that obtained from heterologous tissues. (b) Liver RNA is the best competitor for an annealing reaction between DNA and RNA produced by liver chromatin in vitro. (c) Kidney RNA is the best competitor for an annealing reaction between DNA and RNA produced by kidney chromatin. (d) RNA synthesized by whole chromatin, dehistoned chromatin, and deproteinized chromatin as described by the kinetics of annealing of the synthesized RNA to DNA. (From J. Paul and R. S. Gilmour. 1968. J. Mol. Biol. **34**:305–316. Fig. 5a.)

## Hybridization

Another way of placing chromosomes in foreign cytoplasm is by hybridizing different species. In general, hybrid individuals do not develop successfully, although noteworthy exceptions exist among mammals, such as the mule hybrid between horse and donkey. Hybrids are also occasionally produced between lions and tigers and between several species of bears. Various species of birds can also be hybridized. In the case of the successful hybrids, one must assume that no species specificity of regulatory molecules, such as chromosomal proteins, exists or that any differences are too small to interfere with the normal functioning of chromosomes of either parental genome. Nevertheless, most hybrids fail to develop, and this is particularly true for hybrids among amphibians. In general, hybrids between various species of the genus *Rana* begin to develop but cease development at gastrulation and after a few hours die. One view of this hybrid arrest of development postulates an incompatibility between the two genomes of the parental species. However, this possibility has been completely excluded in recent years by the techniques of nuclear transplantation.

## Nuclear transplantation

The evidence is clear that nuclear behavior is substantially dependent upon the cytoplasm. It also seems probable, although not yet proved, that the nucleus, under proper conditions, is capable of a complete repertory of responses to appropriately programmed cytoplasmic signals. That is, nearly all nuclei from the same individual should be able to reach an identical state of function in the same cytoplasm. The experimental foundations for such a sweeping conclusion are by no means adequate. The most pertinent experiments involve nuclear transplantation. This task with amphibian eggs was first begun by implanting nuclei of early embryonic cells, particularly nuclei from blastula and very early gastrula cells, into enucleated eggs of the frog, *Rana pipiens* (Fig. 8-5). The recipient eggs cleave and not uncommonly give rise to normal embryos. However, when the transplanted nuclei—that is, the donor nuclei—are taken from progressively older cells, the fraction of success in inducing normal development declines. In fact, in the case of *Rana pipiens,* completely normal development has not been obtained from any transplanted nucleus from adult or nearly adult tissue. In the related amphibian *Xenopus laevis,* it has proved possible to obtain normal development using nuclei derived from the intestinal epithelium of larvae. However, even these cells, although apparently mature and functional, are still not from an adult organism (Fig. 8-6). More recent experiments have used nuclei from adult *Rana pipiens* suffering from a Lucké carcinoma of the kidney. Nuclei from these carcinoma cells transplanted into enucleated *Rana pipiens* eggs were capable of inducing development of abnormal tadpoles. These abnormal tadpoles contained many kinds of cells and tissues that appeared normal on cytological examination. Thus, the

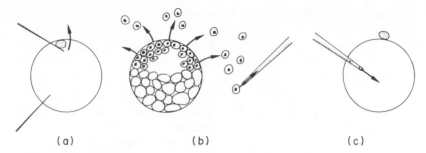

(a)                     (b)                     (c)

Fig. 8-5 The technique of nuclear transplantation in frogs. (a) An unfertilized egg is activated by pricking it with a sharp glass needle (lower left). The maternal genetic material is then removed with another needle, which is moved swiftly in the direction of the arrow (top). (b) A donor embryo, in this case a blastula, is dissociated into single cells by versene, and a single cell is sucked into the transplantation pipette. In there, the cell breaks, baring its nucleus. (c) Now the nucleus, and some cytoplasm, is injected into the enucleated egg which thereafter contains a foreign, diploid nucleus. The exovate which contains its own maternal genetic material (top) later disintegrates. (From H. Ursprung. 1965. *In* Cooke [ed.] The biologic basis of pediatric practice, pp. 1391–1393. Fig. 141-5. Used with permission of McGraw-Hill Book Company.)

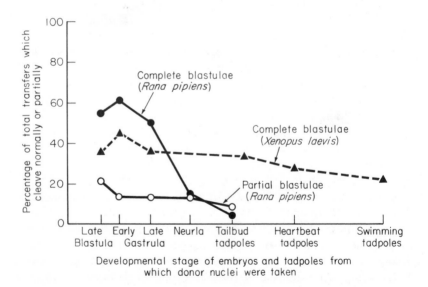

Developmental stage of embryos and tadpoles from which donor nuclei were taken

Fig. 8-6 Graph showing declining capacity of older nuclei to promote normal development after transplantation. (After J. B. Gurdon. 1963. Quart. Rev. Biol. **38**:54–78.)

genes required for producing these many diverse types of cellular differentiation were still present and intact in the nuclei of these tumor cells.

The failure to produce completely normal animals may have resulted from some defective organization of the genome, so that a properly programmed activation of genes could not occur, or perhaps technical difficulties in the transplantation procedure were responsible. The ability of a small nucleus to respond to the large volume of egg cytoplasm has also not been clarified. During embryonic development there seems to be a progressive nuclear differentiation that becomes ever more difficult to reverse, even when the nuclei are introduced into the cytoplasm of the egg. The most likely explanation at the present time for inability of adult nuclei to substitute completely for the egg nucleus probably is the difficulty of quickly erasing the consequences of differentiation of the chromosomes in the transplanted nucleus. The genome of a transplanted nucleus may not be fully exposed to the activating influences that would induce it to carry out normal physiological activities appropriate for the early stages of development.

## Nuclear transplantation between species

By the techniques just described, it is also possible to implant the diploid nucleus of one species into the enucleated egg of another. Such eggs develop up to gastrulation and then cease development at about the same stage as would the embryo of a true hybrid. This failure to develop is a reflection of some incompatibility between the genome of one species and the cytoplasm of another. Moreover, such arrested embryos at the gastrula stage can provide nuclei that may be retransplanted into freshly enucleated eggs. Such second generation transplants again develop to gastrulation and arrest as before. This procedure can be repeated many times with the same results.

Thus, the nuclei of one species are able to "live" and to divide repeatedly in the cytoplasm of a foreign species, but they are unable to provide the basis for development past the gastrula stage. Cytological observation of the arrested gastrula cells frequently shows the presence of gross chromosomal abnormalities in the form of broken chromosomes, rings, minutes (small chromosome fragments), and deletions (Fig. 8-7). These abnormalities are sufficient to explain the failure to develop; however, they still indicate an abnormal type of interaction between the chromosomes and the cytoplasm. At least one hybrid between species of *Rana* is successful. It occurs between *Rana palustris* and *Rana pipiens*. Hybrids of these two species develop to normal adults, exhibiting characteristics intermediate between the two parental species. However, if the nucleus of a *Rana pipiens* cell is transplanted into

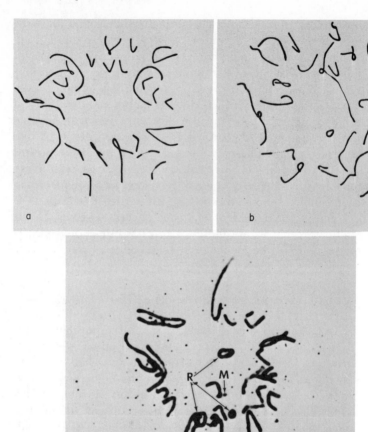

Fig. 8-7   (a) Camera lucida drawing of a metaphase from a control clonal donor. The normal diploid number of 26 chromosomes is present (b) Camera lucida drawing of a metaphase from a nuclear transfer hybrid which was arrested as a late blastula. There are 25 chromosomes including a small ring. (c) Metaphase from an arrested back-transfer gastrula showing rings (R) and a minute (M). Small black dots are pigment granules. (Courtesy of Sally Hennen.)

the enucleated *Rana palustris* egg, development will not proceed to a normal adult but will cease during an early larval stage. Cytological examination of the cells of such an arrested individual do not reveal chromosomal abnormalities, and, furthermore, nuclei from such nucleocytoplasmic hybrids are able to promote normal development when transplanted back into their own cytoplasm. This experiment clearly indicates that the chromosomes of the one species are not irreversibly damaged by residence in foreign cytoplasm, although they are unable to sponsor normal development in it.

Related hybridization experiments have also been conducted in sea urchins. Sea urchins are favorable material for the study of nucleo-cytoplasmic interactions because cross fertilization is feasible in several species. One of the most interesting results of such crosses is the observation that the offspring of reciprocal crosses do not reach the same developmental stage. If, for example, eggs of the sea urchin *Sphaere-chinus granularis* are fertilized by sperm of *Paracentrotus lividus*, perfectly normal pluteus larvae are obtained. Their skeletons are intermediate in appearance between those of the parental organisms. In the reciprocal cross—that is, eggs of *Paracentrotus* fertilized with sperm of *Sphaerechinus*—development of the hybrid offspring is arrested at the gastrula stage. This failure to develop normally is probably attributable to extensive chromosome loss. During the first three cleavage divisions of this hybrid, 16 to 17 of the 20 paternal chromosomes are eliminated; they remain on the equatorial plates and fail to complete the anaphase movements. Later in development, these chromosomes are extruded into the blastocoel. In the opposite cross, which leads to normal hybrid plutei, no such chromosome loss occurs.

The observation of unequal behavior of offspring of reciprocal crosses is not restricted to sea urchins, but has also been observed in amphibians. It is a strong indication that chromosome function depends on the quality of cytoplasm. These various experiments of hybridization clearly indicate that the chromosomes of one species do, in fact, interact with the cytoplasm of the alternate species. Just which molecules are involved and what the nature of the interaction may be have not been elucidated. However, it seems clear that there must be species specificity in the molecular events involved.

## Hybridization of cells in vitro

The failure of amphibian hybrids to develop past the gastrula stage must be contrasted with the dramatic success achieved by hybridizing cells in tissue culture. The cells of various mammalian species, including rat and mouse, man and mouse, and many others, have been successfully fused into single hybrid cells in vitro. The cells of these species are more unrelated than those of the frogs that failed to hybridize successfully. Yet hybrid cells between a mouse and a rat do form; the cells continue to multiply in tissue culture, and enzymatic analysis of these cells proves that at least some genes from both parental species function at the same time in the same hybrid cell. Two enzymes have been examined in detail: lactate dehydrogenase and beta glucuronidase. The behavior of both enzymes indicates a true cell hybrid, but the results for LDH are the more dramatic and are indicated in Fig. 8-8. It can be seen that the LDH isozyme patterns of both the mouse and the rat are evident in the hybrid cells. Lactate dehydrogenase subunits produced by the **A** gene are different for the mouse and the rat. Both genes function and produce subunits that combine at random to make the expected hybrid molecules. This result clearly indicates that the population of cells analyzed could not give rise to this result unless individual cells contained functioning genes from both the rat and the mouse. Subunits for LDH must be available at the same time within a single cell for hybrid tetrameric molecules to form.

In these experiments, the investigator depends on relatively few hybrid cells that form spontaneously. More recently, techniques have been developed that vastly increase the yield of hybrid formation by adding inactivated *Sendai* virus to the co-incubated cell lines. With this technique, such hybrid cells were made between chicken erythrocytes and *Hela* cells. The erythrocyte nuclei began to synthesize nucleic acid,

Fig. 8-8   Zymograms showing rat and mouse LDH isozyme patterns and LDH isozymes synthesized by rat-mouse hybrid cells.

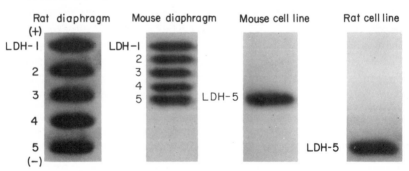

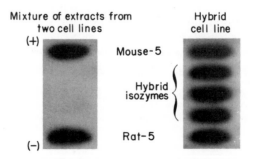

both DNA and RNA, even though they were previously quite inactive (Fig. 8-9). This remarkable technique of cell fusion is effective even for cells from different species or orders of vertebrates. The resulting interspecific hybrid cells survive for very long periods of time, in some cases dividing and establishing new strains of truly hybrid cells.

How may we explain the vastly different results from the successful hybridization of cells in tissue culture as compared with the failure of hybrid development among amphibians, or among mammals, for that matter? What appears to be involved is not simply the mechanisms that turn genes on and off but rather the fastidious requirements for sequential programming of gene function during embryonic develop-ment. Any aberration in the normal duration or amount of gene func-tion during embryonic development could easily disrupt essential sequences of reactions and lead to a termination of development. This is to be contrasted with the requirements for cells in tissue culture. In such cells, only the necessary metabolic reactions for sustaining life need be continued; no progressive change to an alternate state is re-quired. Thus, species-specific mechanisms for activating genes may function in the same cell, but the progressive programming of gene function appears to rest on a degree of species specificity that pre-cludes the successful interaction of the DNA of one genotype with the cytoplasmic molecules of an alternate genotype. Since the hybrid block is a block to development and not to cell viability, we might search for reactions that are critical for development but appear to be un-related to the normal functioning of cells. Such reactions might be those concerning the nature of the cell membrane. This seems reason-able since the failure to develop in hybrid amphibians occurs at pre-cisely the time when changes in the relative association of cells are required to bring about the morphogenetic movements of gastrulation. Remember, however, that mosaic individuals made by the surgical addition of tissues from two different species of amphibians are able to cooperate and carry out the movements of gastrulation. Thus, any species specificity of cell membranes does not preclude cooperative effort between cells from different species. Any incompatibility must therefore lie in the establishment of the properties of the cell membrane itself.

## Meiosis and mitosis

The heterochromatization of the X chromosome in female mammals, or the paternal set of chromosomes in the male mealy bug, demonstrates selective control over certain chromosomes as opposed to others. However, all sexually reproducing metazoans exhibit an even

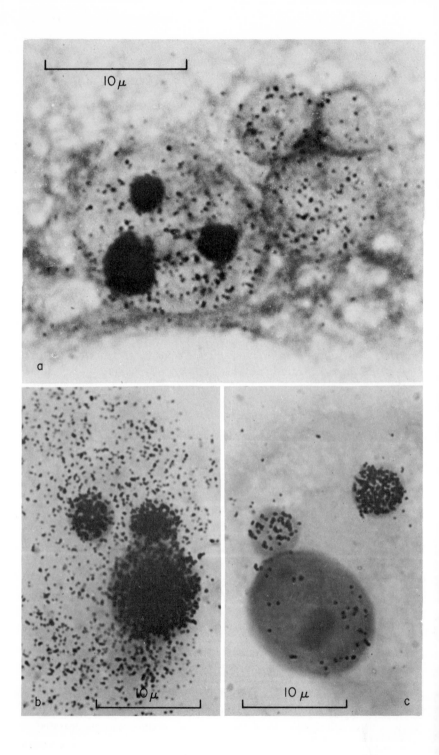

140

more general regulation of gross chromosomal behavior that is apparent in the distinction between meiotic and mitotic cell division. Gametes are derived from somatic cells with the diploid complement of chromosomes. At the termination of a sequence of mitotic replications, the cells enter meiosis, which is characterized by pairing of homologous chromosomes and by delayed replication of the centromeres. These two events must be under genetic control, but the molecular events giving expression to this control are unknown.

An insight into the mechanisms controlling meiosis might be obtained by an examination of the events occurring during the development of parthenogenetic eggs. Many different animals reproduce parthenogenetically—that is, without sexual reproduction—and the eggs of such animals develop without the intervention of sperm. The cytological basis for parthenogenesis is highly varied, depending upon the species concerned, but frequently involves suppression of meiosis at some stage. If this suppression occurs at the beginning of gametogenesis, then cellular differentiation results in diploid gametes with a genetic makeup identical to the maternal organism. The suppression of meiosis can occur at several succeeding stages also. The variety of steps in gametogenesis at which meiosis may be suppressed indicates the multiplicity of molecular events that are probably involved. Several genes must be responsible for the specific molecular events that distinguish meiosis from mitosis, since meiosis is clearly under genetic control and may be interrupted at several stages. The identification of the corresponding gene products provides an area for significant but difficult research. It should be emphasized that chromosome pairing and delayed centromere division are not unique to meiosis. Somatic pairing of chromosomes and somatic crossing-over in *Drosophila* cells is well known, and mitotic cell division involves the timing of centromere division. In the case of meiosis, centromere division is merely delayed. Thus, the molecular events of meiosis do not appear to be unique but only programmed differently from mitosis.

---

Fig. 8-9 Photograph of hybrid cells with included nuclei synthesizing RNA. (a) Autoradiograph of a heterokaryon containing 1 HeLa nucleus and 3 reactivated erythrocyte nuclei. The cell was exposed to tritiated uridine for 20 min. If the nucleolar labeling in the HeLa nucleus is neglected, the total number of grains over the three erythrocyte nuclei together is not very different from the number of grains over the HeLa nucleus. (b) Autoradiograph of a heterokaryon containing 1 A9 nucleus and 2 reactivated erythrocyte nuclei which have not yet developed nucleoli. The cell was exposed for 6 hr to tritiated uridine. All the nuclei are heavily labelled and there is substantial labelling of the RNA in the cytoplasm. (c) Autoradiograph of a heterokaryon similar to the one shown in (b) and at the same stage of development. The cell was also exposed to tritiated uridine for 6 hr, but after irradiation of the A9 nucleus with an ultraviolet microbeam. The erythrocyte nuclei continue to synthesize RNA, but there is very little labeling of the RNA in the cytoplasm. (Courtesy of H. Harris.)

A search for the distinctive events that lead the cell into meiosis would logically fall in the period of prolonged prophase characteristic of meiotic cells. Several biochemical and cytochemical studies have been made of this period of cell development, and, as one might expect, both RNA and protein synthesis occur, but specific macromolecules have not been identified. Nevertheless, the synthesis of these macromolecules is clearly essential to the completion of meiosis, as shown by the failure of meiosis to proceed to completion when protein synthesis is inhibited. In a study of meiosis in the lily, a significant difference in DNA synthesis was found to distinguish meiotic from mitotic cells. During meiosis, a delayed synthesis of a very small fraction of nuclear DNA, about 0.3%, occurs. This DNA is distinctive in terms of its density and base composition and may be involved in chromosome pairing, but the manner of its involvement is quite obscure. No evidence at all has yet been uncovered to suggest a molecular basis for that other distinctive aspect of meiosis—namely, late centromere duplication. Nevertheless, these two varieties of the cell cycle, meiosis and mitosis, further illustrate the sensitive interdependence of chromosomal and cytoplasmic events in regulating cell behavior.

## Maternal effects

Examples of nucleocytoplasmic interactions are also provided by progeny of reciprocal crosses that do not coincide in phenotype. The offspring of a cross between a female horse and a male donkey differ in weight and stature from the offspring of the reciprocal cross— that is, a female donkey and a male horse. In both cases, the offspring resemble the mother more than the father. Such cases have been termed *maternal effects* or *maternal inheritance*. In the cases cited, it appears that the uterine environment influences the development of the offspring.

In suitable experimental organisms, it has been possible to reveal the mechanisms operating in maternal inheritance. In *Drosophila*, at least one example is known of an enzyme that functions during the early periods of development and is entirely of maternal origin. This conclusion was made possible by the use of structural variants of the enzyme aldehyde oxidase—variants that affect the electrophoretic mobility of this enzyme. Figure 8-10 shows the result of a gel electrophoretic analysis of aldehyde oxidase prepared from two species of *Drosophila*. Notice that in crosses of the two species, the enzyme present during early development is always that electrophoretic variety of

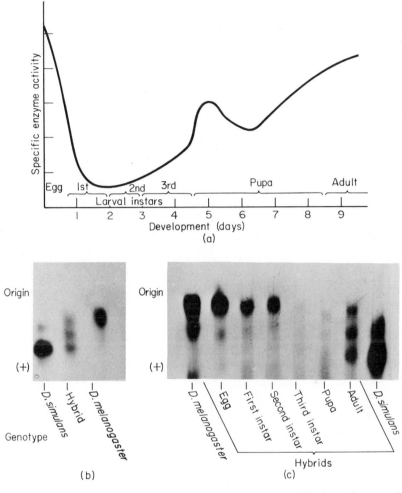

Fig. 8-10 Maternal effect in *Drosophila* aldehyde oxidase. (a) Aldehyde oxidase specific activity is high in the egg, and then drops rapidly, to rise only slowly during the larval stages. In order to test the hypothesis that the egg aldehyde oxidase was of maternal origin, a cross (b) was carried out between flies with aldehyde oxidase differing in electrophoretic mobility. Notice in (c) that the offspring indeed contains mostly maternal enzyme in early stages. In the adult, maternal, paternal, and also a hybrid enzyme are present.

aldehyde oxidase present in the maternal organism. Only at later stages of development does the paternal form of the enzyme appear in the hybrid organism. It is not known whether the maternal organism passes intact enzyme molecules into the developing egg cells or whether, alternatively, long-lived messenger RNA containing the genetic information required for the production of aldehyde oxidase is synthesized by the oocyte nucleus and stored in the egg.

In the case of a different *Drosophila* enzyme, xanthine dehydrogenase, it is known that maternal transmission of functional enzymes cannot be responsible for the maternal effect typical for that particular enzyme. We have described the various mutants involved in the genetic control of xanthine dehydrogenase elsewhere (p. 68). When a female, heterozygous for the ma-1 gene, is crossed to a male that is hemizygous for the same mutations, one would expect to find wild-type and mutant flies in equal numbers among the progeny of this cross. This is not the case. Rather, all the progeny of this cross are phenotypically wild type—that is, they resemble their mother. Of course, it would appear possible that the maternal organism passes intact xanthine dehydrogenase molecules into the egg cell. That this is not so was shown unambiguously by a cross diagrammed in Fig. 8-11. In this cross, a fly heterozygous for the maroonlike gene and homozygous for the rosy gene was used as the maternal organism. Recall that rosy is the structural gene for xanthine dehydrogenase. An organism homo-

Fig. 8-11  Diagram of *Drosophila* cross to demonstrate maternal effect on xanthine dehydrogenase.

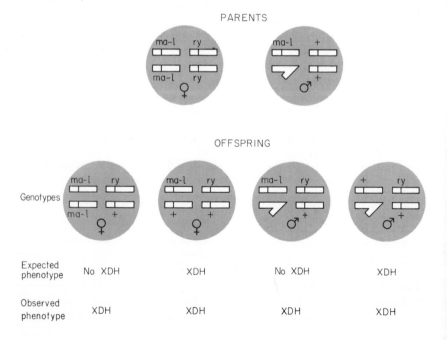

zygous for the rosy mutant allele does not synthesize xanthine dehydrogenase. Consequently, the maternal organisms used in our cross could not transmit finished enzyme molecules to the progeny. Just what *is* transmitted in the case of this maternal effect is not known.

Toward the end of the last century, the view was expressed that the cytoplasm, not only the nucleus, contains informational material that can be replicated and is passed on to daughter cells in cell division. More recently, it was found that all cells examined contain DNA in the cytoplasm, mostly, if not exclusively, in mitochondria and chloroplasts. This DNA is remarkably similar in a wide variety of organisms; it exists as a circular molecule 5 µ long, and accordingly has a molecular weight of several million. Mitochondrial DNA can replicate itself and transcribe RNA, which, in turn, is used in protein synthesis. It would not be surprising if mitochondrial DNA were indeed the carrier molecule of cytoplasmic inheritance of some maternal effects, but little information is currently available on this question.

## REFERENCES

Bolund, L., N. R. Ringertz and H. Harris. 1969. Changes in the cytochemical properties of erythrocyte nuclei reactivated by cell fusion. J. Cell Sci. 4:71–87.

Briggs, Robert and Thomas J. King. 1957. Changes in the nuclei of differentiating endoderm cells as revealed by nuclear transplantation. J. Morphol. 100:269–312.

Brown, S. W. 1966. Heterochromatin. Science 151:417–425.

Brown, Spencer W. and Uzi Nur. 1964. Heterochromatic chromosomes in the coccids. Science 145:130–136.

Goldstein, Lester and David M. Prescott. 1966. Protein interactions between nucleus and cytoplasm. *In* Lester Goldstein [ed.] The control of nuclear activity. Prentice-Hall, Inc., Englewood Cliffs, N.J.

Gurdon, J. B. 1963. Nuclear transplantation in Amphibia and the importance of stable nuclear changes in promoting cellular differentiation. Quart. Rev. Biol. 38:54–78.

Hennen, Sally. 1963. Chromosomal and embryological analyses of nuclear changes occurring in embryos derived from transfers of nuclei between *Rana pipiens* and *Rana sylvatica*. Develop. Biol. 6:133–183.

Hennen, Sally. 1965. Nucleocytoplasmic hybrids between *Rana pipiens* and *Rana palustris*. I. Analysis of the developmental properties of the nuclei by means of nuclear transplantation. Develop. Biol. 11:243–267.

Lyon, M. 1966. X-chromosome inactivation in mammals. *In* D. E. Woollam [ed.] Advances in Teratology 1:25–54. Logos Press, distributed by Academic Press, New York.

Lyon, M. 1968. Chromosomal and subchromosomal inactivation. Ann. Rev. Genetics 2:31–52.

Ursprung, H., K. D. Smith, W. H. Sofer and D. T. Sullivan. 1968. Assay systems for the study of gene function. Science 160:1075–1081.

Weiss, Mary C. and Boris Ephrussi. 1966. Studies of interspecific (rat × mouse) somatic hybrids. II. Lactate dehydrogenase and β-glucuronidase. Genetics 54:1111–1122.

# NINE

## Genetic Interactions in Cell Differentiation

Genes are clearly important, even decisive, in determining the course of cell differentiation and in fixing the properties of adult cells. Although the primary function of each gene may be simple, the orchestration of many genes in the quantity and timing of their function to yield the properties of the differentiated cell is exceedingly complicated. We may gain a deeper appreciation of the role of genes in controlling cell differentiation by analyzing the appearance of a characteristic "gene product" during the differentiation of a specific type of cell. One gene product that has been much studied by geneticists and embryologists is the pigment melanin, exclusively synthesized in melanocytes. This pigment is the product of a sequence of biochemical reactions in which the enzyme tyrosinase plays an essential role (Fig. 9-1). In mammals, this enzyme is under the control of a single gene at the C (color) locus. One mutant variety of this gene fails to form detectable tyrosinase. Individuals homozygous for this mutant gene cannot synthesize melanin and are albinos. However, a great many different genes in addition to the structural gene for tyrosinase are known to affect the differentiation of melanocytes.

### Melanocyte differentiation

Vertebrate organisms differentiate at least two kinds of melanocytes (Fig. 9-2). One type is the epithelial melanocyte, which develops in the proximal half of the optic vesicle and forms the pigmented

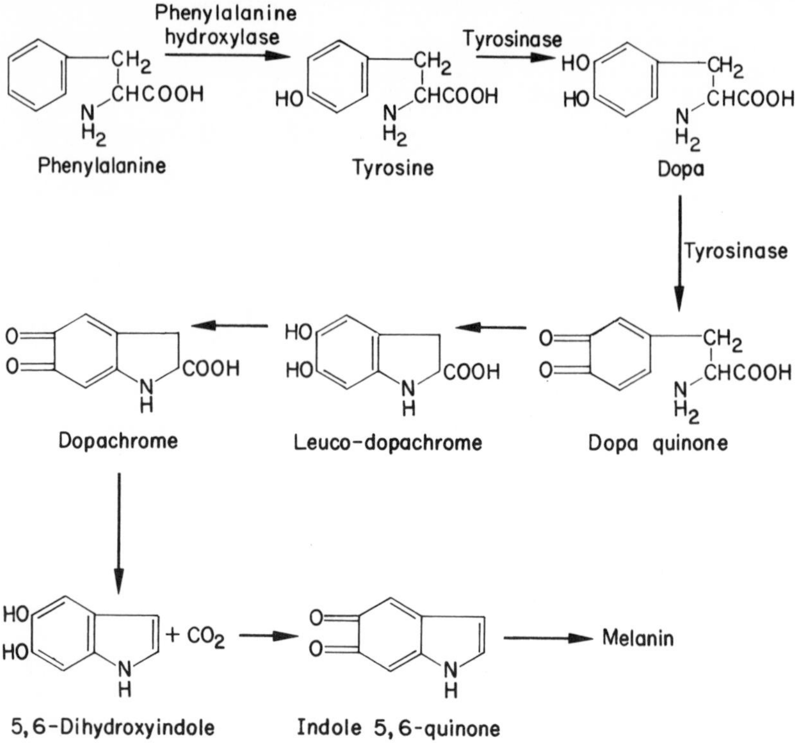

Fig. 9-1   Biochemical pathway of melanin synthesis.

retina of the mature eye. The second type of melanocyte arises from the neural crest as a dendritic melanoblast cell that migrates into many different tissues, such as skin, hair follicles, and the choroid coat of the eye. In each of these tissues, it may complete its differentiation to become a fully pigmented melanocyte. During the early stages of differentiation, including the period of migration, the melanoblasts produce neither tyrosinase nor melanin pigment; consequently, these embryonic cells are difficult to identify. Cell multiplication must occur while the cells are still melanoblasts, because mitotic figures are seldom seen in mature, fully pigmented melanocytes. Differentiation in this cell line is subject to genetic control at every step. The number and distribution of the melanocytes, the morphology of the cell and its characteristic organelle—the melanosome, the color of the melanin, and the release of pigmented melanosomes from the cell are all influenced by identified genes, some of which act primarily within the

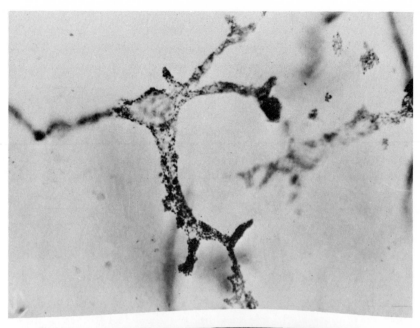

Fig. 9-2  Photograph showing two kinds of melanocytes; dendritic and epithelial. (Top) Dendritic melanocyte from mouse harderian gland. (Bottom) Dense layer of epithelial melanocytes composing the pigmented retina of the mouse eye. Layer indicated by arrow. Pigmented layer next to retina is the pigmented choroid coat composed of densely packed dendritic melanocytes.

melanoblast whereas others function within the surrounding cells that make up the tissue environment.

The formation of melanosomes by melanoblasts signals the terminal step in the differentiation of melanocytes. The melanosomes are complex cell organelles composed of subunits of fibrillar protein that is assembled within membrane-limited vesicles to form sheetlike matrices in which the enzyme tyrosinase is normally embedded. Melanin is then deposited upon these protein structures until the melanosome is fully pigmented and enzymatic activity is lost. The organization of the melanosomal matrix and the pattern of deposition of melanin within the melanosome are under the control of several different genes (Fig. 9-3). One gene in mice affecting melanosome structure is the gene (p) for pink-eyed dilution. The presence of this gene in homozygous form (pp) results in a disorganized matrix owing to the failure of the melanosome fibrils to cross-link properly. Consequently, the melanosomes of pink-eyed individuals are not heavily pigmented. When incubated in the presence of tyrosine in vitro, these melanosomes do produce additional melanin, but the defect in matrix structure is not corrected. Nevertheless, one effect of the presence of this gene is to diminish the overall pigmentation of the fur.

Another gene that indirectly affects the total apparent pigmentation of the fur of mammals is the gene **d,** for dilution. This gene does not directly affect the amount of pigment formed but does determine the shape of the melanocyte. Melanocytes of the **dd** genotype have a greatly reduced number of dendrites and an enlarged cell body. These melanocytes, with their stubby dendrites, transfer fewer melanosomes to the surrounding keratinocytes, which are destined to form the hair. In addition, these melanocytes are not so well anchored in the hair follicle and frequently break loose from their normal locations to become incorporated into the developing shaft of the hair. This results in large clumps of melanosomes rather than in a uniform dispersion of granules throughout the keratinocytes. Such a distribution of melanosomes gives the superficial appearance of diluted pigmentation, although the total pigment present has not been reduced.

Environmental influences of a nongenetic nature also may influence the behavior of melanocytes to yield variations in the amount of pigment formed. Mice of the genotype $c^h c^h$ have mutant alleles of the structural gene for tyrosinase and produce an altered thermolabile form of the enzyme. Mice of this genotype produce much less pigment

---

Fig. 9-3    EM photograph of genetically different melanosomes, albino (a), pink-eye (b), black and brown (c). The albino melanosome has been magnified to about 4 × the size of the other melanosomes. (Courtesy of Frank Moyer.)

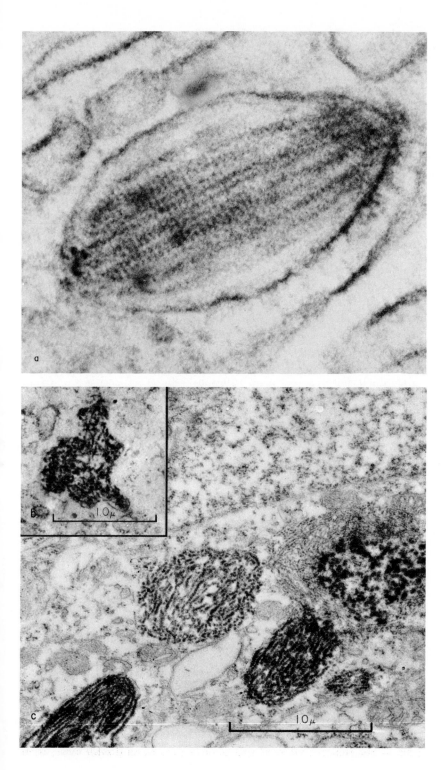

at warm temperatures than at relatively cold temperatures (10°C). Thus, temperature affects the phenotypic expression of these particular alleles at the **C** locus. It is also well known that melanoblasts respond to ultraviolet light. The tanning process in man as a result of exposure to sunlight appears to involve an increase in both the number of pigmented melanocytes and the number of melanosomes per cell. Whether the ultraviolet stimulus to increased melanization acts directly within the melanocyte or in the surrounding cellular environment has not been clarified.

The pigmentation of many animals changes with the season. Some animals have a white pelage in the winter and black or brown pelage in the summer. These changes are mediated through a variety of hormones, such as gonadotropins, corticotropin, and melanocyte-stimulating hormone. Even in human beings, hormonal changes associated with pregnancy result in increased pigmentation of various areas of the body, such as the nipples and the areolae. But, again, it is not known whether these hormones act directly on the melanocyte to alter pigment formation or whether they affect primarily the cells of the tissue environment that, in turn, affect the melanocyte. However, since the hormones apparently have equal access to all the pigment cells, which nevertheless react differently in accord with the tissue environment, it seems reasonable to conclude that the action of these hormones is mediated via the cells of the tissue environment.

Melanocytes do not normally divide, and their supply must constantly be replenished through differentiation from a pool of melanoblasts. The melanoblasts do undergo mitotic replication, but with increasing age the supply of melanoblasts diminishes, as is evident in the white hair of aging human beings. Alternatively, the aging environment of the melanoblasts may simply lose the capacity to induce melanocyte differentiation. The white hairs of an old individual contain many cells that are not keratinocytes and that resemble unpigmented melanocytes in some respects. However, no melanosomes are apparent in these cells, and if their differentiation was indeed halted by age-related changes in the cellular environment, then the cessation of development must have occurred at a very early stage. In any event, the genetic control of pigmentation, whether acting within the melanocyte or through the tissue environment, is modified by the changing conditions brought about by age. This is perhaps simply another expression of the fact that the function of the cell's genome is sensitively dependent upon the state of cell differentiation and is also the major contributor to the state of differentiation.

Of all the genes that affect melanocyte differentiation, the most fundamental is the structural gene for the enzyme tyrosinase, the **C** gene. If the inactive allele of this gene is present in double dose, then an albino individual results. Such individuals contain both kinds of melanocytes, epithelial and dendritic, but these cells are unable to complete their differentiation. No tyrosinase is synthesized and no melanin pigment forms, but all the preceding steps in differentiation do occur to yield a "pigmentless melanocyte." Such cells can be recognized by their characteristic morphology and location, as well as by the presence within the cell of melanosomes, although without their usual pigment. Albino melanosomes are essentially normal in structure, as shown by observations with the electron microscope (Fig. 9-3). The genes for melanin formation, of course, act in both the epithelial and dendritic melanocytes. Thus, albino individuals have pink eyes and white skins as well as white fur or feathers. Clearly, the gene for tyrosinase is essential for the complete differentiation of melanocytes. However, this gene appears to be totally unimportant for the differentiation of any other kind of cell. All other cells in albino individuals appear to be normal. As might be expected, tyrosinase can be detected only in melanocytes. The simplest conclusion is that this gene is permanently inactive in all the cells of the body except those that normally make melanin.

The melanocyte has many identifiable characteristics, including the ability to form different colors of melanin. These characteristics are, of course, influenced by the activity of genes. Again, some of these genes are active within the melanocyte itself, and others are active in cells that provide an environment for the melanocyte. Most instructive are the nine allelic forms of the gene at the **A** locus. Arranged in order of dominance of expression, these alleles are indicated in Fig. 9-4. Various alleles at the **A** locus control the presence or absence of yellow

| | |
|---|---|
| $A^y$ | lethal yellow |
| $A^{vy}$ | viable yellow |
| $A^w$ | light-bellied agouti |
| $A^d$ | dark back, light-bellied agouti |
| A | agouti |
| $a^{td}$ | tanoid |
| $a^t$ | black and tan |
| a | non-agouti |
| $a^e$ | extreme non-agouti |

Fig. 9-4  Allelic series of genes at the agouti locus. Agouti pigmentation produces a kind of salt and pepper type of effect due to the fact that a black agouti hair has a subterminal band of yellow pigment a short distance below the tip of the hair. In yellow genotypes the amount of yellow pigment is extended. In black genotypes the yellow pigment is reduced or absent.

pigment. A mouse carrying two **a** alleles is described as nonagouti and its fur is black. In such animals, all melanocytes, wherever located within the animal, produce black melanin. At the opposite end of the allelic series is the allele for yellow pigment, **A**$^y$. This allele, when present in heterozygous form, **A**$^y$**a**, leads to an almost uniformly yellow fur color. However, in such mice, those melanocytes located outside the hair follicles—for example, in the retina of the eye and in the skin—continue to produce black pigment. Only the melanocytes exposed to the environment provided by the hair follicle produce yellow pigment. Thus, these mice synthesize two kinds of pigment, the particular kind produced depending upon the tissue location of the melanocyte.

The relative roles of melanocyte genotype and the cellular environment in bringing about melanocyte differentiation have been dramatically demonstrated by skin transplantation experiments between mice of different genotypes (Fig. 9-5). In mature skin, melanoblasts do not normally migrate very far from the locations they reached during the embryonic life of the mouse. However, at birth, the structure of the mouse skin is still somewhat labile, and disruption of the tissue structure associated with transplanting bits of skin enables melanoblasts in the disrupted area to migrate to hair follicles at some distance from their original location. Thus, when a small piece of skin from a pigmented animal is transplanted to an albino host shortly after birth, melanoblasts from the pigmented skin migrate into the hair follicles of the albino host and there give rise to pigment. By such grafting procedures, melanocytes of yellow genotype (**A**$^y$**a**) can be grafted into non-

Fig. 9-5 Drawing showing design of skin transplantation experiments between mice of different pigment genotypes.

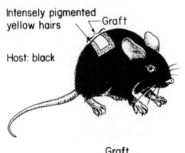

Intensely pigmented yellow hairs — Graft

Host: black

Graft
nonpigmented but
carries genetic determinants
for yellow

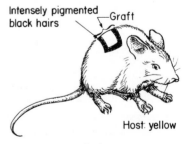

Intensely pigmented black hairs — Graft

Host: yellow

Graft
nonpigmented but
carries genetic determinants
for black

agouti (**aa**) animals. The transplanted **A$^y$a** melanocytes migrate out of the grafted skin and into adjacent host tissues, where they populate hair follicles and eventually produce black pigment, despite their own genotype. The reciprocal transplantation experiment leads to reciprocal results—that is, melanocytes of **aa** genotype will migrate out of donor black skin into surrounding hair follicles of **A$^y$a** hosts and there give rise to yellow pigment. Those donor melanocytes that migrate into tissues outside hair follicles continue to form black pigment even though adjacent cells are **A$^y$a** in genotype. These transplantation experiments clearly demonstrate that the responsibility for the formation of yellow pigment does not rest upon the genes of the melanocyte but rather upon the genes of the hair follicle cells. Hair follicle cells of the **A$^y$a** genotype apparently elaborate a stimulus that provokes the melanocyte to make yellow pigment. In the absence of this guiding stimulus, the melanocyte always makes black pigment. Thus, genes at the **A** locus, so important in the differentiation of the melanocyte, function effectively only in hair follicle cells. These genes appear to be silent in all other cells, including the melanocyte itself.

Mouse melanocytes are able to form not only yellow and black pigment but also brown pigment. The decision as to whether a cell will form brown or black pigment is controlled by a series of alleles at the **B** locus—namely, **B$^{lt}$** (light), **B** (black), **b$^c$** (cordovan), and **b** (brown). If the alleles for brown (**bb**) are present in melanocytes, they will elaborate brown pigment. The **BB** genotype gives rise to black pigment. However, the effects of these alleles are subordinate to instructions from follicle cells to manufacture yellow pigment. Such yellow pigment appears to be the same in cells of **BB** or **bb** genotype. When the melanocyte receives instructions to form yellow pigment, it does so if it is able to make any pigment at all—that is, if it contains a gene (**C**) for color. If the formation of yellow pigment involves enzymatic activity by the melanocyte beyond the enzymatic activity involved in the formation of black or brown pigment, then one might anticipate the discovery of mutations that would affect the capacity of a melanocyte to form yellow pigment. However, no such mutant has been reported. Since a vast amount of observation and research has been directed toward the pigment genes of mammals, we may tentatively conclude that the formation of yellow pigment does not require enzymatic activity in the melanocyte beyond that required to make black pigment. Reasoning in this fashion leads a developmental geneticist to conclude that yellow pigment is formed as the result of a follicular cell product entering a melanocyte and there transforming the color of the incipient pigment from black or brown to yellow.

An additional, more subtle example of the dependence of gene action on the state of cell differentiation is exemplified by the behavior of cells carrying the $a^t$ allele at the **A** locus. This allele, described as black and tan, is responsible for the elaboration of black pigment in the hair on the back and tan or yellow pigment in the hair on the belly. Thus, the hair follicles on the dorsum and the hair follicles on the ventrum elaborate different instructions for the melanocytes, even though the genetic makeup of all the cells is identical—namely, $a^t a^t$. Investigators are presently unable to detect any difference between dorsum and ventrum hair follicle cells, other than that provided by such biological tests as the differential effect of the $a^t$ allele. Yet the course of differentiation of these follicle cells in different areas of the skin must be different. Hair follicle cells on the ventrum are induced by the black-and-tan gene to elaborate the stimulus for yellow pigment formation, whereas the corresponding follicle cells on the dorsum behave as if they contained the alleles for non-agouti (**aa**). This example is particularly significant for future analyses, because it indicates the degree of specificity that may be achieved by the interaction of cells of distinct phenotype, although of the same genotype, in guiding the course of cell differentiation.

Apparently identical cells, characterized by somewhat different embryonic histories during the development of the organism, frequently exhibit different kinds of gene expression, as is so dramatically illustrated by mouse cells carrying the black-and-tan alleles. As stated earlier, many different tissues of mice contain pigment cells. Among these tissues is the Harderian gland of the eye. Melanoblasts infiltrate this gland and differentiate into normal dendritic melanocytes. However, several genes, in mutant form, prevent the differentiation of melanocytes in harderian glands (Fig. 9-6). Some of these genes have little or no effect on the formation of pigment in other areas of the body. Thus, we are forced to conclude that the presence of the mutant alleles of these particular genes in the harderian gland is responsible for the elaboration of a somewhat altered environment by these harderian cells, so that melanoblasts in this tissue are unable to complete their differentiation.

The most dramatic example of alternative forms of melanocyte behavior in similar tissue environments is presented by the white spotted pigment patterns so characteristic of many mammals and birds. White spotted animals obviously develop melanocytes that are fully capable of forming melanin in many areas of the body, but they are unable to do so in the white spotted areas. The cells of the white areas apparently fail to permit the entrance, survival, or the differentiation of melanocytes. Cytological observation of the hair follicles in the white spots reveals no early stages in the differentiation of melanocytes,

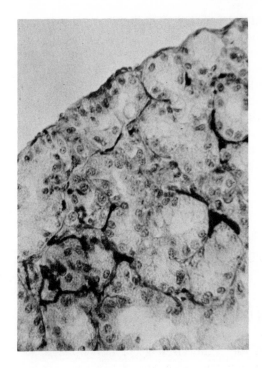

Fig. 9-6 Photograph of harderian gland showing infiltration of melanocytes.

which is contrary to what is true for albino individuals. Albinos, although white throughout their tissues, do contain recognizable melanocytes of normal morphology, but they are devoid of melanin. However, melanoblasts that enter white spotted areas early in development apparently do not complete those stages required to make them morphologically recognizable. Alternatively, the white areas may have proved lethal to embryonic melanoblasts. However, it is possible by skin transplantation shortly after birth to encourage melanoblasts from pigmented tissue to enter white spotted areas. Melanoblasts from the transplanted skin that are in a relatively advanced stage of differentiation migrate into the adjacent white spotted regions, enter the hair follicles, and there make pigment. Thus, the white spotted areas are capable of sustaining the differentiation of melanoblasts after they have passed some early critical step in differentiation—a step that cannot be supported by the skin of a white spotted area.

These few cases, and there are many more, illustrate the complexity and sensitivity of interaction between the differentiating cell, its environment, and the genetic makeup of both the responding and inducing cells. It is the interaction of all these variables that leads to the extraordinary specificity of differentiation of individual cells at particular locations in the animal. Quite obviously, the terminal characteristics of the cell are attributable to the activity of its genes plus the substances available from the environment. The most fundamental question concerns the molecular responsibility for selectively activating genes and for maintaining all others in a quiescent state. So far, the discussion has proceeded as if genes were either on or off. However, the expression of gene activity ranges over a wide spectrum from barely perceptible to grossly conspicuous, which is also true for the genes regulating melanin synthesis.

## Allophenic mice

Recently, a remarkable advance in experimental technique has made available unusually valuable experimental material for studying the role of genes in cellular differentiation. This material is in the form of allophenic mice.

Allophenic mice are genetic mosaics arising through the combination of blastomeres from embryos of two different genotypes. These mice are readily constructed in the laboratory by skillful investigators. First, embryos at about the eight-cell stage of development are isolated from the oviducts of pregnant mice, and the blastomeres are dissociated with the potent proteolytic enzyme pronase. Then the

Fig. 9-7    Diagram of experimental procedure used to produce allophenic mice and allophenic mice showing striped pigment patterns. (After B. Mintz. 1969. *In* Bergsma and McKusick [eds.] First conference on the clinical delineation of birth defects. Birth Defects: Original Article Series. New York. The National Foundation. 5:11–22.)

### Allophenic mice from aggregated eggs

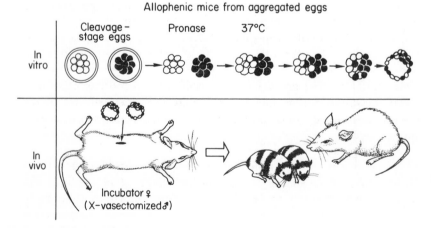

blastomeres from two or more embryos may be recombined (Fig. 9-7), and, when firmly attached to one another, may be reimplanted in a properly prepared female mouse. Development will commonly continue and eventuate in the birth of a mosaic mouse, that in effect has been produced by four or more different parents. These mice provide valuable insights into the role of genes in the differentiation of cells and in the regulation of the interaction among cells during embryonic development. Such allophenic mice have already provided answers to several important questions in developmental genetics. First, it has long been known that the striated muscle cells of the body contain many nuclei, but mitosis is never observed in these mature cells. How do these multinucleated muscle cells arise? They might originate through repeated nuclear division in myoblasts (embryonic muscle cells) without cytoplasmic division, or they might arise through the fusion of mononucleated myoblast cells. A clear decision between these alternatives has been provided by allophenic mice: striated muscle cells arise from the fusion of mononucleated cells. This conclusion was demonstrated by allophenic mice composed of cells from two different genotypes with reference to the isozymic forms of isocitrate dehydrogenase (IDH). This enzyme, like the previously analyzed lactate dehydrogenase, is polymeric. The allelic forms of the responsible gene encode electrophoretically distinguishable subunits of the enzyme. Heterozygous individuals produce three forms of this enzyme, the intermediate form representing the hybrid molecule composed of subunits stemming from both alleles. When blastomeres from the two alternate homozygotes are combined, they give rise to allophenic mice that differentiate muscle cells which synthesize all three forms of IDH (Fig. 9-8). The hybrid IDH could only have been produced in cells containing both allelic genes functioning at the same time. Thus, cells from each genotype must have fused to form single multinucleated muscle cells containing nuclei of both genotypes.

A second problem of great significance, not yet completely elucidated even by allophenic mice, is the mechanism of immunochemical tolerance. The developmental origin of the immune system is still obscure, although the area is under active investigation and significant advances have recently been made. Several different, highly differentiated cell types appear to be involved in the normal immune response to foreign antigens and to foreign grafted cells. The cooperative action of the cells responsible for the immune response results in a highly selective rejection of foreign cells and foreign antigens. Allophenic mice can be constructed of cells from antigenically incompatible individuals, as demonstrated by rejection of grafts between the adults. However, such allophenic mice, composed of immunochemically incompatible cells, nevertheless develop normally, and their diverse cells

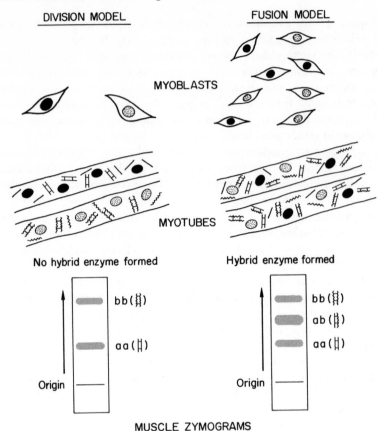

MUSCLE ZYMOGRAMS

Fig. 9-8 IDH zymogram from allophenic mice showing hybrid molecules formed in fused muscle cells. (From Mintz and Baker. 1967. Proc. Nat. Acad. Sci. U.S. 58:593, Fig. 1.)

function harmoniously without any evidence of immune conflict. The general immunochemical capacities of such mice are not suppressed, inasmuch as they exhibit normal rejection responses to new antigens encountered after birth. Thus, juxtaposition in allophenic mice of genetically unlike cells prior to the differentiation of the immune system precludes any subsequent rejection of either cell type. Although two different genotypes are involved, the cells nevertheless recognize one another as not foreign, even though each cell continues to produce its characteristic antigenic products. These mice are intrinsically tolerant of all the different antigens produced by their two different cell types. Thus, immune tolerance is substantially a developmental phenomenon rather than a simple direct expression of an animal's genotype.

A more subtle problem in developmental genetics is also illuminated by allophenic mice: fixing the topographic relationships among cells in the adult animal as a consequence of their interactions during embryonic development. The pigment patterns of allophenic mice from parents of different pigment genotypes show clearly that the total pattern emerges from the repeated cell division of melanoblasts located at a restricted number of different positions along the anterior-posterior axis of the developing embryo. Each of these positions is apparently occupied exclusively by one melanoblast at an early stage in development. Once the position is occupied, all other melanoblasts are excluded. In allophenic mice, the adult pigment pattern will reflect the characteristics of the two different kinds of melanoblasts, in accord with the random distribution of these two genotypes to each location. The pigment pattern can be a mosaic of alternating stripes when the primordial melanoblasts of each genotype happen to alternate along the anterior-posterior axis of the early embryo (Fig. 9-7).

Studies of allophenic mice have already provided new and sometimes startling insights into several old problems in genetics and development. We may be confident that these fascinating mice will continue to provide us with a rich store of data with which we can analyze genetic and cellular interactions during embryonic development.

REFERENCES

Billingham, R. E. and W. K. Silvers. 1960. The melanocytes of mammals. Quart. Rev. Biol. **35**:1–40.

Markert, Clement L. and W. K. Silvers. 1956. The effects of genotype and cell environment on melanoblast differentiation in the house mouse. Genetics **41**:429–450.

Mintz, B. 1969. Developmental mechanisms found in allophenic mice with sex chromosomal and pigmentary mosaicism. *In* D. Bergsma and V. McKusick [eds.] First conference on the clinical delineation of birth defects. Birth Defects: Original Article Series. New York, The National Foundation. **5**:11–22.

Mintz, B. and W. W. Baker. 1967. Normal mammalian muscle differentiation and gene control of isocitrate dehydrogenase synthesis. Proc. Nat. Acad. Sci. U.S. **58**:592–598.

Mintz, B. and W. K. Silvers. 1967. "Intrinsic" immunological tolerance in allophenic mice. Science **158**:1484–1487.

Moyer, Frank H. 1966. Genetic variations in the fine structure and ontogeny of mouse melanin granules. Am. Zool. **6**:43–66.

Silvers, Willys K. 1961. Genes and the pigment cells of mammals. Science **134**:368–373.

# TEN

# Genes and
# Morphogenesis

## Pleiotropism

The earlier literature of developmental genetics is full of examples of genes affecting morphogenesis. Mutants have been found that induce the formation of an abnormal tail in the mouse, that lead to cyclopia or anencephaly in guinea pigs, to the lack of limbs in calves, and to deformed beaks or the absence of kidneys in chickens, to mention but a few examples. The study of the sequential developmental steps that lead to such gross abnormalities is very complicated and seldom leads to a clear explanation. Nevertheless, such research is of great importance in view of the frightening number of genetically controlled congenital malformations in man. These range from very harmless malformations, such as webbing of toes (Fig. 10-1), to more serious malformations, such as phocomelia (Fig. 10-2). Congenital malformations affecting almost every developmental event have been observed in humans, and we shall not understand these abnormalities completely until we understand normal development completely.

So little is presently known about the genetic control of mammalian development that we cannot in the near future hope to understand the paths that must lead from the altered gene to the complex abnormal phenotypes. Nevertheless, many of the complex syndromes have been traced to simpler events closer to the primary function of the gene. An informative analysis is that of the mouse mutation, congenital hydrocephalus (ch). This syndrome consists in an abnormal number

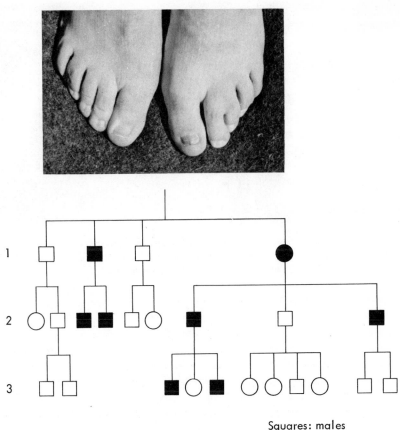

Squares: males
Circles: females
Open symbols: normal
Black: with webbed toes

Fig. 10-1   The trait "webbed toes" in man. The webbing, particularly noticeable between toes 2/3 on the left foot, does not involve bones, but only soft parts. It is clearly under the control of a dominant, autosomal gene as evident from the pedigree.

and arrangement of whiskers, failure of the eyelids to be closed in the embryo, abnormal shape of the pituitary gland, absence of dermal flat bones of the skull, and cerebral hemorrhage, which leads to the death of these mice immediately after birth. The bulging foreheads of these newborn mice are filled with hemorrhagic fluid, and the animals have blunt snouts. From the genetic information available, a single Mendelian unit is responsible for this complex syndrome. This is not an exceptional case. As a rule, genetic defects that manifest themselves at

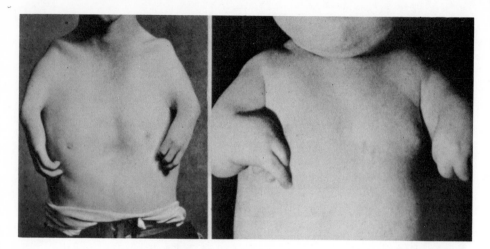

Fig. 10-2    Genetically determined (left) and thalidomide induced (right) phocomelia in man. (From H. Ursprung. 1965. *In* Cooke [ed.] The biologic basis of pediatric practice, pp. 1391–1393. Fig. 141-3. Used with permission of McGraw-Hill Book Company.)

a morphological level are pleiotropic in nature—that is, they cause a multitude of effects. And yet many of these multiple effects can be traced back through development and reduced to a few, and occasionally to one, primary genetic effect. For the case of the **ch** mouse, such a "pedigree of causes" has been elaborated (Fig. 10-3), which makes hydrocephalus directly responsible for the entire group of symptoms. Hydrocephalus itself is brought about by a growth anomaly in the cartilaginous skull so that the skull is shortened. The brain is forced upward, thereby impeding the proper drainage of cerebrospinal fluid. When the cartilage cells forming the cartilaginous skull were

Fig. 10-3    Pedigree of causes in the congenital hydrocephalus syndrome of the mouse. (From H. Ursprung. Fig. 141-1.)

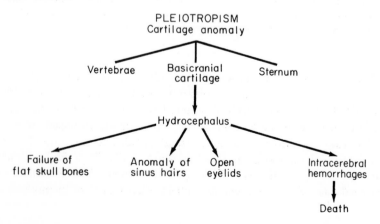

examined histologically, they proved to be abnormal, with extensive vacuolization and a highly abnormal matrix. This cartilage anomaly is not restricted to the skull, but is also present in the vertebrae, the sternum, and other parts of the skeleton. The common denominator of all the symptoms is the cartilage anomaly. It leads to a fundamental disturbance in morphogenesis, which, in turn, is responsible for the appearance of many of the secondary and tertiary phenotypic effects.

The causal analysis of pleiotropism has been carried somewhat farther in a few mutants, such as the pituitary dwarf mouse (**dw**). The two major symptoms of this syndrome are dwarfism and sterility. At the histological level, the thyroid and adrenal cortex are found to be hypoplastic, and a careful investigation of the pituitary gland shows that its anterior lobe is retarded in its growth. The pituitary of the mutant mouse contains only small cells with pycnotic nuclei, but lacks the large acidophils that, in normal animals, produce growth hormone. From the viewpoint of basic research, this observation is not very satisfying, because we have no idea just how the **dw** allele affects the proper differentiation of acidophils. However, the discovery has proved to be very important for therapy, for it gives the investigator the knowledge needed to correct the developmental abnormality. If the primary defect is, indeed, an effect on hormone production by acidophils of the pituitary, then administering normal pituitary hormone should correct some or all of the secondary and tertiary symptoms. This therapy was attempted on dwarf mice. The mice received daily implants of wild-type pituitaries from the day of birth. As Fig. 10-4 shows, they reached normal size and, in fact, grew to be sexually mature individuals. The thyroid gland and adrenal cortex also became normal in size during this treatment. The pituitary itself did not become normal, however. This result supports the assumption that the pituitary cells are those primarily affected by the mutant locus.

## Phenocopy

Tracing a developmental abnormality back in ontogeny to its earliest expression offers the hope of identifying the primary genetic effect. From such an identification, one might expect to rectify at least the consequences of the defective gene. A different approach consists in attempting to imitate a genetically caused abnormality by interfering with normal development through the use of physical or chemical treatments. Such experimentally produced abnormalities frequently appear almost indistinguishable from those caused by genetic changes. We have pictured a genetic case of phocomelia in Fig. 10-2; compare

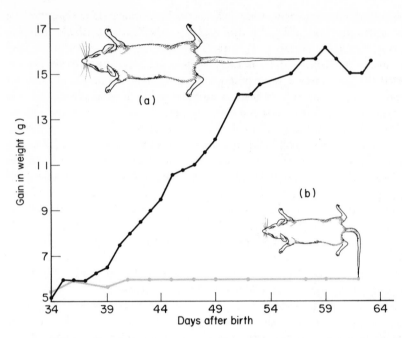

Fig. 10-4　Effect of implantation of normal pituitaries into *dwarf* mice on their body weight. Normal pituitaries were implanted into *dw/dw* mice daily, starting 34 days after birth (a). In the absence of this treatment, the mice remained very small (b). (From E. Hadorn. 1955. Letalfaktoren, p. 237–285. Fig. 128.)

this with the thalidomide-induced abnormality shown in the same figure.

A hormone therapy analogous in logic to the one just described for the dwarf mouse is thyroxine treatment of cretinism in humans. Several causes are known for cretinism, some of which are inherited. L-thyroxine, given daily to the defective child starting on the first day of life, has proved quite effective in preventing severe mental retardation. In one series of patients, 12 of 29 cretins that had been adequately treated during the first year of life had IQ's greater than 90. (No patient, among 50 treated *after* the first year of life had an IQ greater than 90; thus, early diagnosis and treatment are essential.)

Such phenocopies of genetic defects have been produced by many different experimental treatments. The terminal effects of the gene *rumplessness* in a chicken, for example (Fig. 10-5), can be imitated by injection of insulin into the yolk sac at an early stage of development. Interestingly enough, the same compound, when injected at a later developmental stage, leads to the imitation of a different genetic lesion, short upper beak (Fig. 10-5). The different organ primordia of de-

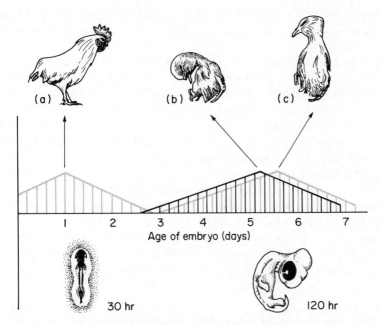

Fig. 10-5 Phenocopies induced by insulin in the chick. The ordinate indicates the relative sensitivity as a function of developmental stage (abscissa). If insulin is given during days 0–3, rumplessness is induced. If it is given around day 5, short upper beak is produced. The lower figures show stages of development representative of these two critical phases. (From E. Hadorn. Fig. 101.)

veloping organisms are sensitive to phenocopying agents at particular phases of development. These phases have been called phenocritical phases, and similar critical phases have also been postulated for gene action. For example, hybrid arrest occurs at specific developmental stages that are not randomly distributed over the life span of an organism. Time specificity of gene action would be expected to generate critical stages of development.

Many investigators have used the phenocopy approach in searching for the primary genetic lesion in congenital abnormalities. They assume that similar effects may have similar causes. Insulin-produced abnormalities are similar to those produced by the mutants short upper beak or rumplessness; therefore, these genes may induce metabolic disturbances similar to those produced by excessive insulin administration. Initially, much support for this notion came from genetic experiments. Modifier genes were discovered which, when built into the genome of mutants, affected the expression of the mutant phenotype. These same modifier genes also affected the expression of the phenocopying substances. But the phenocopy approach suffered a blow when it was found that very different agents produced the same malformations; boric acid, for example, was found to produce the same terminal effects as insulin.

The fact that a multitude of different causes can produce the same phenocopy is well documented in a number of experimental systems. In *Drosophila,* for example, both ether and high temperature, applied at an early embryonic stage of development, produce flies that look just like tetraptera mutants (Fig. 10-6). (Tetraptera flies have four wings, two on the mesothorax and two in the most posterior segment of the thorax, which in the mutant is not a metathorax, but rather another mesothorax.)

It is not difficult to understand how a variety of substances can produce the same end effect. Developmental processes are long chains of events, many of which are under genetic control, but many of which are specified by the cellular environment. It does not matter very much at which point the chain of events is disturbed; the end result will be compounded of similar defects, no matter where the initial defect was introduced into the chain of events.

Fig. 10-6.   Ether phenocopy of tetraptera in *Drosophila.* The ordinate indicates the yield of phenocopies resulting from ether $(C_2H_5)_2O$ (black curve) or heat treatment (gray line) given at various developmental stages (abscissa). The drawing on the upper left shows a section through the blastoderm, at which stage the embryo is most susceptible to the induction of the tetraptera phenocopy. (From E. Hadorn. Fig. 99.)

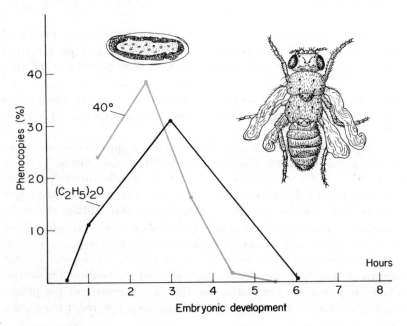

## Embryonic induction and morphogenesis

Developmental processes are often a result of interactions between different cell types in the embryo. We have previously mentioned the role of hormones in such interactions. Two kinds of genetic interference with such interactions are to be expected: first, genes could control the production and release of the hormonal signal needed for a developmental process; second, genes could control the ability of a cell to respond to the hormonal stimulus.

The hormonal circuitry controlling insect development has been intensively studied. At least two hormones are directly responsible for growth and differentiation in insects. One of these, the growth and differentiation hormone, controls cell division and enables the cell to undergo specialization. The second, the juvenile hormone, determines whether a cell remains in a larval state or enters metamorphosis. In *Drosophila,* these two hormones are produced by a small organ called the ring gland. The identification of this gland was made possible because of a mutant gene, lethal giant larva, which prevents metamorphosis of the larva. This failure to metamorphose suggested that the mutant gene might control the appropriate hormonal stimulus. In a search for the organ that normally produces and releases this hormonal stimulus, a variety of wild-type organs was transplanted into lethal giant larvae. Implanted wild-type ring glands, but no other tissues, initiated metamorphosis in the mutant larvae. Apparently the cells of the hypodermis responsible for the formation of the puparium have the inherent capacity to function even in the abnormal genotype, but the ring gland of the mutant does not elaborate the hormonal signal required to trigger the initiation of puparium formation.

Other cell types of lethal giant larvae also maintained the ability to respond to a normal hormonal stimulus, as shown by a reciprocal experiment. The larval primordia of ovaries, which in lethal giant larvae never develop, were transplanted to wild-type hosts. In these hosts, the transplanted ovaries differentiated in synchrony with the ovaries of the host; at least the somatic components of the gonads differentiated. The germ cells did not complete differentiation. Thus, two cell types, ring gland cells and germ cells, appear to be directly affected by the mutation. Imaginal disc cells—that is, those primordia giving rise to the adult fly during metamorphosis—also are affected by this mutation; these primordia degenerate during early larval life. On the other hand, the intestine, nervous system, and the larval muscles are all normal in the mutant. Other tissues, such as the fat body and the salivary glands, are present in the larva, although much reduced in size (Fig. 10-7). The example of lethal giant larva illustrates that release of a hormonal signal and the ability to respond to the signal can both be under genetic control.

Many other genes controlling cell and tissue interactions have been extensively investigated. The Danforth short tail mutant of the mouse is particularly instructive. In mice of this genotype, kidneys often fail to develop. During normal development, the formation of the kidney appears to be triggered by an inductive interaction between the ureters and the metanephrogenic mesenchyme. A histological study of these mice demonstrated that the mutant ureters fail to grow sufficiently to make contact with the mesenchyme. Consequently, no inductive interaction occurs. Another case of a mutant gene interfering with inductive interaction during development has been described in guinea pigs. Embryological experimentation in a variety of vertebrates has repeatedly shown that the mesoderm in the roof of the archenteron induces, in the overlying ectoderm, the formation of neural structures. Experimental disturbance of the archenteron is followed by a variety of abnormalities, particularly in the head. Anencephaly, microcephaly, and cyclopia are commonly produced. Several mutant genes in guinea pigs induce head malformations that resemble those produced experi-

Fig. 10-7 Induction of puparium formation in *lethal giant* (lgl) *Drosophila* larvae by implantation of normal ring-glands. The ordinate indicates the percentage of animals that have completed puparium formation at the time indicated on the abscissa. The normal control animals pupate rapidly. The lgl animals with implanted wild-type ring glands develop more slowly but much faster than the lgl larvae with no ring glands implanted. (From E. Hadorn. Fig. 125.)

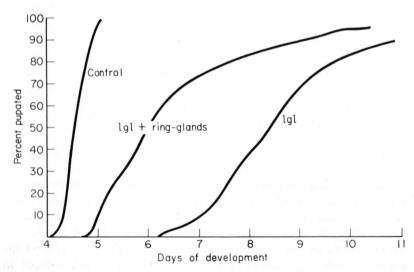

mentally. These mutations affect solely the head regions of guinea pigs, producing a series of abnormalities ranging from an almost normal appearing skull to extreme anencephaly (Fig. 10-8). It is not known how the mutant genes produce this developmental disturbance, but it is evident that the cells normally giving rise to the skull fail to proliferate as a consequence of the genetic lesion.

In other deviant cells, the genetic constitution permits proliferation, but the cells differentiate abnormally. We will discuss examples of this type of genetic interference with normal development in the section on neoplastic transformations (Chap. 11).

## Imaginal discs

In our discussion of phenocopies, we have encountered the four-winged mutant, tetraptera, of *Drosophila*. The underlying mutations are called *homoeotic,* meaning that they confer the properties of one body segment on another. Tetraptera is not the only homoeotic mutant. Tetraltera is another, so named because it carries four balancers (halteres) on the thorax instead of the normal two wings and two halteres. Aristapedia is a third homoeotic mutant, characterized by the transformation of antennae into legs. Recently a new

Fig. 10-8 Otocephaly in the guinea pig. Notice the wide range of expressivity, ranging from normal morphology (10) to severe deformation of the skull (12). (From E. Hadorn. Fig. 122.)

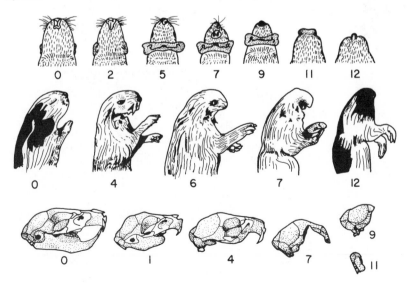

homoeotic mutant was discovered and named *Nasobemia,* after a fictitious animal called "Nasobem" in a poem by the German poet Christian Morgenstern. This fictitious animal walks on its nose. The mutant *Nasobemia* grows a leg from its palpus!

To understand the ontogeny of homoeotic mutants, we must be familiar with insect development. Flies have a complicated life cycle. About one day after fertilization, a larva, the first instar larva, hatches. After another day, this larva, which has grown considerably, sheds its old cuticle and emerges as the second instar larva. The second instar larva grows for about a day, molts, and gives rise to a third instar larva. After another two days of feeding and extensive growth, this larva enters pupation. Its cuticle hardens and forms the pupal case or puparium. Inside the puparium a most remarkable process, metamorphosis, starts. In the course of about 4 days, many larval tissues are destroyed and the adult fly, or imago, is built up. Most structures of the imago are pieced together during metamorphosis from the so-called imaginal discs (Fig. 10-9). Imaginal discs are groups of cells,

Fig. 10-9    Development of *Drosophila* imaginal discs.

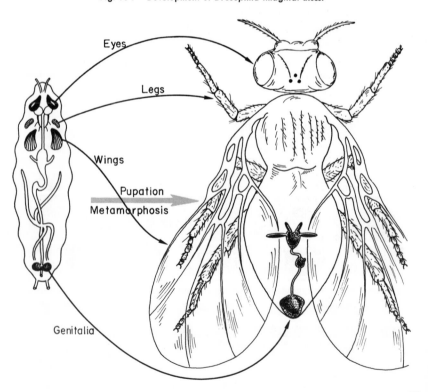

present throughout all the larval stages, but with no apparent larval function. During metamorphosis, they give rise to imaginal or adult structures. For example, each leg is formed from a leg imaginal disc, each compound eye from an eye disc, each wing from a wing disc, each haltere from a haltere disc, and so forth. The head capsule is formed by a joint effort of several discs, including the eye discs. The genitalia are formed from the genital disc, which is located just anterior to the larval anus.

Histologically, imaginal discs consist of layers of rather uniform epithelial cells arranged around a narrow lumen. The cells of these discs show little sign of differentiation; they all look very much alike. They have large nuclei and relatively little cytoplasm, and are filled with free ribosomes. Nevertheless, the evidence indicates that the cells of the discs are rather rigidly programmed to proceed down certain developmental pathways. This can be shown at various levels. If an entire wing disc is removed from one larva and injected into the body cavity of another larva, it will form a wing when the host larva undergoes metamorphosis (Fig. 10-10). It will form this wing regardless of its location within the host. Likewise, an eye will form, autonomously, when an eye disc is implanted into the body cavity of a host. The larva from which a wing disc has been surgically removed will metamorphose into an adult lacking one wing. Thus, discs are rigidly determined and have the capacity for self-differentiation in a foreign environment.

The determination of the disc goes even farther. Figure 10-11 shows

Fig. 10-10 Transplantation of imaginal discs. Imaginal discs are dissected free from a donor larva (b). After metamorphosis (M), the implants are visible through the body wall of the host fly (c, dotted arrow) and can be dissected free, since they usually float within the body cavity.

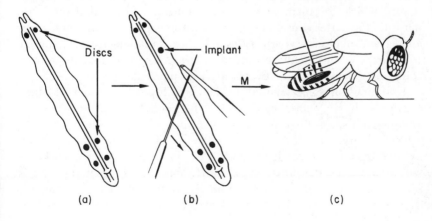

(a)          (b)          (c)

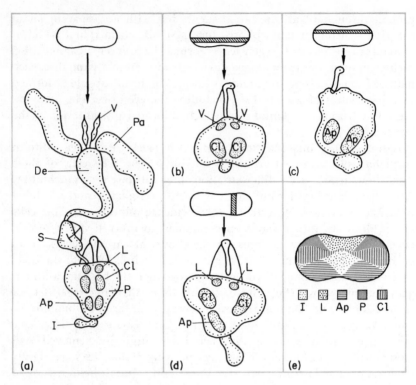

Fig. 10-11    The fate map of a genital imaginal disc as revealed by UV microbeam irradiation. (From H. Ursprung. 1963. Am. Zool. 3:71–86. Fig. 1.)

the developmental capacities of a genital imaginal disc. Several parts of the genitalia form from this disc: two vasa deferentia, which become hooked up with the gonads; two accessory glands called paragonia; and a contractile ductus that transports sperm into the sperm pump, which, in turn, moves sperm into the external genital apparatus. The latter consists of several chitinous parts, including a penis and several plates carrying bristles and thorns that engage with similar structures on the female fly during mating. The two anal plates surrounding the anus and the two claspers are particularly noticeable.

Two lines of experimental evidence indicate that each of these organs is already represented within the genital disc at an early developmental stage. First, if pieces cut from a disc are each cultured in a larva, then each will give rise, predominantly, to only one structure. Second, exposure of small parts of a disc to an ultraviolet microbeam will prevent the development of individual organs when the disc is reimplanted into a host larva. From the results of such experiments, fate maps have been drawn for several imaginal discs, including the genital disc and the wing disc (Fig. 10-11).

## The bithorax series

With this information, it is now possible to understand the phenomenon of homoeotic mutations. Five different mutations (**a, b, c, d, e**) will be discussed, which alone or in combination affect portions of the mesothorax (meso), metathorax (meta), or the first abdominal segment (**abd**). A diagram illustrating the normal configuration of segments in a fly is shown in Fig. 10-12. The mutation **a** causes the anterior portion of meta to develop into a structure that looks like the anterior portion of meso. These animals have two normal wings and two abnormal wings; the latter appear to be formed by a transformation of the anterior portion of the haltere disc into anterior wing parts. This mutant is called *bithorax*. At the other end of the spectrum, the mutation **e** *(postbithorax)* causes the posterior end of meta to re-

Fig. 10-12 Normal configuration of *Drosophila* appendages. (From E. B. Lewis. 1963. Am. Zool. 3:33–56. Figs. 3, 5a.)

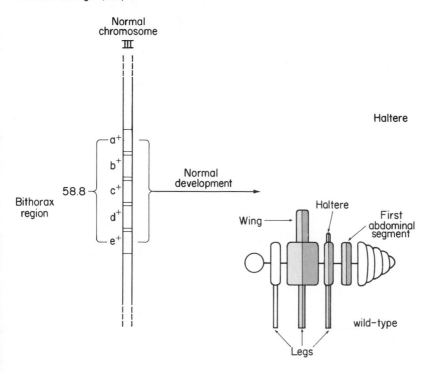

semble the posterior end of meso. When a cross is made to produce a fly that contains both **a** and **e** mutations, the entire meta is transformed into meso; the fly thus has a double mesothorax and is completely four winged (Fig. 10-13).

The **d** gene is similar in effect to the **e** gene, but goes farther in causing **abd** to look just like meta of an **e** animal. The abdominal segment of such flies may bear halteres, and the abdominal segment now has legs. The fly thus has eight legs instead of the six characteristic of all insects. Usually, **c** animals are lethal when homozygous, but some viable flies have been found carrying a mutation at this locus, and these showed all the transformations that we have listed thus far. Finally, in **b** animals, the posterior portion of meso is altered to resemble the posterior portion of meta. The diagram shown in Fig. 10-13 summarizes the various effects attributable to these five mutations.

Fig. 10-13    Developmental consequences of homoeotic mutations. (From E. B. Lewis. Figs. 3, 5b, d, 6a, 8a.)

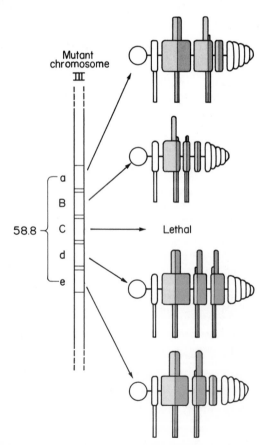

This phenomenon is significant for understanding both genetics and development. From the viewpoint of genetics, note that these genes are very closely linked (Fig. 10-14). They are clustered at locus 58.8 on the third chromosome. Genetic recombination analyses demonstrated that these five genes formed a cluster with very rare crossing-over, namely a recombination frequency of approximately 1:10,000 gametes. We have encountered such cases of clusters of related loci before, in the case of the bacterial operon. In *Drosophila,* such clusters are called complex loci, or a pseudoallelic series of genes. Typically, members of such pseudoallelic series show the cis-trans effect—that is, the effect of two mutant pseudoalleles in a single individual depends upon their arrangement in the two homologous chromosomes. The fly of the genotype **a+/+c**, for example, has a mutant appearance, whereas a fly of a genotype **ac/++** resembles the wild type. This cis-trans effect is a form of position effect; clearly, the same genes behave differently when on separate chromosomes. [Since mottling is not observed in the cis-trans position effect, this position effect is called an S-type (stable) effect as opposed to the V-type (variegated) position effect, discussed previously—see p. 71.]

When genetic combinations of the various mutants of this pseudo-

Fig. 10-14 Genetic and cytogenetic organization of the bithorax region of the third chromosome. (From E. B. Lewis. Fig. 9.)

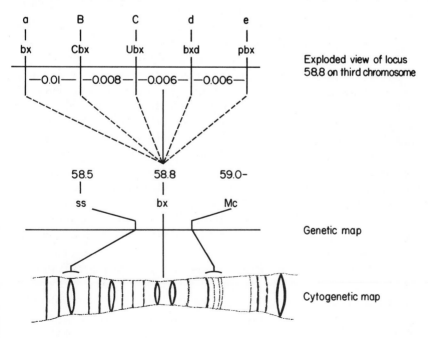

allelic series were made, a conspicuous polarity effect became apparent. For example, a trans heterozygous **d+/+e** does not show a transformation typical of **d** but a transformation typical for **e** (**de/++** is wild type). Flies of the genetic constitution **a+/+e**, on the other hand, show a transformation similar to **e**, but none of the **a** type. The trans heterozygous mutant **c+/+d** shows the **d** transformation. It is evident that the bithorax series shows a polarity in the sense that each locus tends to influence the other loci to the right, but not to the left. Operationally, what happens is that wild-type alleles to the right of a locus are inactivated, permitting the recessive mutant allele to become expressed.

The bithorax series resembles the bacterial operon because of clustering of functionally related genes and because of the polarity effect. Moreover, the products of these genes may be rather directly responsible for the observed gross changes in morphology. Specifically, it has been postulated that the **e+** gene manufactures a substance $s_e$ that suppresses the potential meso-like development of meta. The **d+** gene is thought to produce a different substance $s_d$ that suppresses the potential meta-like development of abd. The **a+** gene would thus produce a substance $s_a$ that suppresses the potential meso-like development of meta. Common to these hypotheses is the assumption that wild-type alleles make a substance that the mutant alleles fail to make. If this were so, dosage effects might be demonstrable for wild-type genes, but not for their mutant alleles. This can be tested by constructing fly genomes that contain one or more doses of the various alleles. Indeed, even several doses of a mutant allele do not yield a mutant phenotype as long as at least one wild-type allele is present in the genome.

The explanation offered is that some gene product is preventing, or suppressing, development. The **d+** allele suppresses the development of an abdominal segment into a thoracic segment. The mutation **d-d+** may be analogous to the mutations that transformed a multi-legged ancestor into the six-legged insects of today. We are assuming here that abdominal segments, indeed, have the potential to form thoracic segments. On the other hand, the mutation **a-a+** would have reduced the four-winged insect to a two-winged insect. In embryological terms, this transformation involves the reduction of the wing disc of the metathorax to a haltere disc. Early in evolution, the developmental potential of segment-forming primordia or imaginal discs must have been very great, but mutations have restricted the developmental capacities of these primordia.

One mechanism by which this could have been achieved is through mutations affecting the capability of some cells in a disc to divide. As we see from the fate map of the wing disc, blocking the proliferation of the posterior part would lead to a b-type animal. This hypothesis would be strongly supported if the presumed dormant cells could be stimulated to proliferate. A disc would then give rise to a structure that it does not normally form. We have already seen that such is not generally the case; otherwise, fate maps could not have been established. But there are a few cases in which a disc has formed structures that it does not normally form. In fact, in one instance, a *Drosophila* disc has formed a structure that is not normally formed anywhere on the *Drosophila*. When a haltere disc is removed from a larva and injected into a host larva, it forms all the structures typical for a haltere, plus a group of adventitious bristles that are not normally found on any body segment of *Drosophila*. The only obvious disturbance of disc development in this case is the transplantation, which in many instances is known to induce some additional mitotic activity. It is therefore conceivable that a cluster of dormant cells was stimulated to proliferate and differentiate, thus giving rise to a new group of structures. This result supports the belief that at least a few cells of an imaginal disc do not differentiate during metamorphosis.

A quite different explanation of homoeosis and the formation of adventitious bristles assumes that the mutation or the surgical interference changes the fate of some cells of the disc. Changing the normal fate would mean acquiring a new kind of determination. This process has been termed *transdetermination,* and very good evidence has been obtained to show that it may indeed operate in the embryological experiments to be discussed now, and possibly also in the genetic experiments just described.

## Transdetermination

Does a programmed cell located in one of the determined areas shown on fate maps of an imaginal disc retain its state of determination when provoked to divide more often than normally? In other words, is determination maintained and transmitted through cell heredity? To answer this question, proliferation must be induced in a piece of imaginal disc for abnormally long periods of time; then the cells must be induced to differentiate to assess the state of determination. Insects are ideal organisms for this kind of experiment. The hormonal milieu of larvae and adults permits imaginal discs to proliferate, but not to differentiate. Only in a pupal environment will cells undergo extensive differentiation. To stimulate proliferation, a disc is removed from its larval donor, cut in two (this is apparently necessary, for an organ that has reached its terminal size will not grow unless injured), and injected into a larva or adult, both of which support proliferation. Proliferation results in a steady increase in size of the implant (Fig. 10-15). Several tissues have been maintained in such in vivo cultures for many years in a number of laboratories. The developmental potentiality of the implants may be tested at any time by transplanting them into larvae. When the larvae metamorphose, the implants differentiate.

In a typical experiment, one-half of a male genital disc was implanted into the abdomen of an adult fly. After 3 weeks, the tissue was removed. It had grown to several times the size of the original implant. This overgrown disc was cut into many fragments, and the fragments were forced through metamorphosis by implantation into larvae ready to pupate. The resulting differentiating structures were recovered from the abdominal cavity of the host flies and examined microscopically. Some of the fragments had developed chitinous structures, such as anal plates or claspers. Others had formed a sperm pump, and still others, paragonia. This behavior is in accord with the fate maps. The experimenter simply happened to isolate individual organ primoria from the tissue mass, and these developed according to their state of determination. This determination can be maintained during long periods of proliferation. In one experiment, anal plate clones continued to give rise to only anal plates during more than 70 transfer generations, lasting several years.

Occasionally, this strict cell heredity of determination is lost. An

Fig. 10-15   Proliferation of imaginal discs in adult host flies.

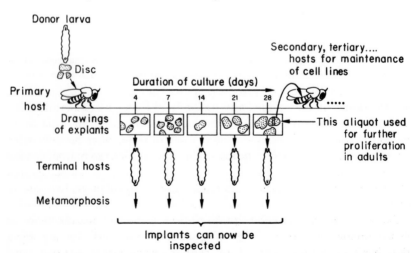

aliquot of anal plate primordium cells can suddenly give rise to a leg, or an antenna, or an eye, for example. Thus, in these cases genital disc tissue differentiates into structures that are not normally formed by the genital disc at all but by a disc of a different body segment. When this occurs, the genital disc has undergone transdetermination. The behavior seems reminiscent of what we observed in homoeotic mutants.

Figure 10-16 is a diagram of the transdeterminations that have been observed thus far. Clearly, cell lines derived from a single imaginal disc can be made to form almost any structure provided that they are kept in culture long enough. Furthermore, there appears to be a definite order in which the various structures appear. The new program of development for a transdetermined structure can be maintained and propagated by cell heredity for many transfer generations. That is, tissue giving rise to legs, although formed from original anal plate material, will remember its new leg program indefinitely. Before we unreservedly accept these results we should rule out some obvious objections to the experimental technique and to the interpretations.

First, do the transdetermined cells originate in the implanted disc, or are they migrant cells from the host organism? This question is easily answered by the use of genetic markers. Routinely, in these experiments the implant differs in genotype from the host. For example, ebony tissue may be cultured in yellow hosts. All host disc cells are, of course, yellow in genotype, and, when differentiated, produce yellow chitinous products. All implanted disc cells on the other hand are genetically ebony, and, when differentiated, are black. Implants, whether transdetermined or not, all show the ebony color.

Second, does the genital disc contain some leg cells that are revealed under the experimental conditions of forced proliferation? This possibility is more difficult to exclude. What is needed are clones derived from single cells previously proved to be determined as anal plate, for

Fig. 10-16   Transdeterminations of imaginal discs. (From E. Hadorn, 1967, Major Problems in Developmental Biology, p. 92.)

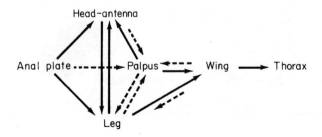

example. If such clones exhibited transdetermination, then the cryptic cell objection would be refuted. Although no one has succeeded in cloning single insect cells, a genetic trick has enabled investigators to examine a clone derived from a single cell. The genetic trick takes advantage of rare instances of somatic crossing-over, as described earlier in the discussion of cell lineage in eye development. Larvae heterozygous for yellow and singed ("yellow" is yellow body color; "singed" is a mutant causing abnormal bristle shape) were X-rayed. When somatic crossing-over occurs between the centromere and the marker genes, two cells differing in genotype result. One is now double homozygous for the two marker genes; the other is double homozygous for their wild-type alleles. The double homozygous mutant cells are recognizable, after differentiation, by the color and shape of bristles. Since somatic crossing-over is a rare event, it occurs only in randomly located cells dispersed throughout the tissue. Since this rare event is equivalent to a mutation, it is inherited by all progeny of the cell in which the event occurred. In a clone labeled this way, both normal and transdetermined structures were found. Therefore, transdetermination cannot be a consequence of selection of a cryptic cell carrying a different program, but must be due to a heritable change in program (Fig. 10-17).

Third, is somatic mutation responsible for transdetermination? At first sight, mutation does seem a very plausible explanation, particularly

Fig. 10-17    Transdetermination occurring in a cell clone. (From W. Gehring. 1968. Results and problems in cell differentiation, vol. 1. Springer-Verlag, Berlin. (From E. Hadorn, 1967, Major Problems in Developmental Biology, p. 92.)

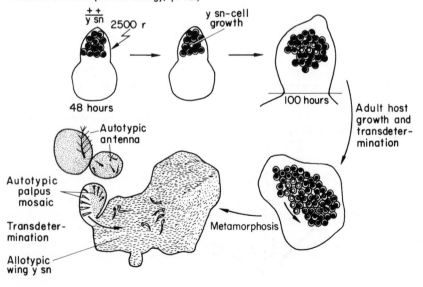

in view of the bithorax series. A transdetermination from wing to haltere would be equivalent to the action of the **b** gene in the bithorax series.

In fact, somatic mutation is difficult to exclude. The frequency of observed transdetermination is orders of magnitude higher than known mutation rates, but we should realize that the frequency of somatic mutation in proliferating disc cultures has not been determined. A more convincing argument against somatic mutation is that transdetermination occurs with a somewhat predictable frequency and in a predictable direction. For example, a thorax forms more frequently from a wing disc than an eye forms from an anal plate disc.

The mechanism of transdetermination is therefore not known. The only known prerequisite for transdetermination is cell proliferation (Fig. 10-18). With so little known, hypotheses are not very restricted. Interpretations of the bithorax series postulated gene-controlled substances that prevented the development of wild-type discs in another direction. Perhaps similar substances become diluted during excessive proliferation, to derepress hidden morphogenetic potentials. This raises a more general question of the stability of the differentiated state, which we shall examine in the following section on regeneration.

**Fig. 10-18.**  The correlation of cell multiplication and transdetermination.

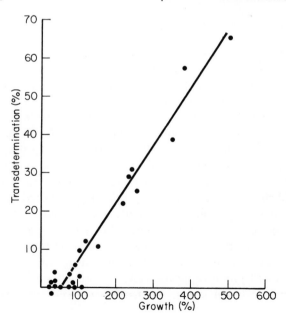

# Regeneration and the programming
# of gene function

The cellular differentiation occurring in normal development is brought about, in part at least, by sequential, selective, gene activation and repression. Whether this effect is achieved at the level of gene amplification, transcription, or translation is not important for the present discussion. What we want to discuss here is whether this sequence of events is irreversibly programmed for a given cell type. We do not think so. In the previous section, we described the loss of determination as a consequence of rapid cell proliferation in imaginal discs. Moreover, a new developmental program was acquired by the transdetermined cells. Is such a change possible once a cell has reached a state of terminal differentiation?

Perhaps the best place to look for examples of reprogramming of gene function is in the regeneration of parts of fully differentiated organisms. Three main questions should be considered. Do cells lose their differentiated state before giving rise to regenerate tissue? Is cell division a prerequisite for participating in regeneration? Can cells during regeneration give rise to new cell types (metaplasia), or do cells always breed true?

## Amphibian limb regeneration

When a limb of an adult newt is amputated, the epidermis soon seals off the wound. After some time, a regeneration blastema forms, not as a simple outgrowth of the amputated limb but as a consequence of reorganization of the limb stump, extending a considerable distance proximal to the site of amputation. It was once believed that "embryonic reserve cells," or fibroblasts, were responsible for this blastema formation. This is clearly not so. Rather, cells of the stump lose their differentiated state, synthesize new DNA, and divide. Electron microscope studies have shown that muscle cells of the stump gradually lose their myofibrils, their most characteristic differentiated feature. The use of tritiated thymidine and autoradiography shows that the nuclei of these same cells begin DNA synthesis. These cells, and not embryonic reserve cells, make up the bulk of the regeneration blastema. In larval newts after limb amputation, the cartilage cells proliferate, after losing the typical cartilaginous appearance. These cells then make up a substantial part of the blastema. Next, blastema cells differentiate into muscle cells, cartilage cells, and connective tissue cells of the regenerating limb. Epidermis, and also blood vessels and nerves, do not form from blastema cells but grow out from the intact tissues left behind in the stump.

It is clear in this case that certain mature cells do lose their differentiated state before giving rise to the regenerate, and that they divide extensively. However, there is no direct evidence that metaplasia occurs. Each cell type in the regenerate may be derived only from antecedent cells of the same type.

## Amphibian lens regeneration

When the lens of an adult salamander eye is removed, a transformation of cells of the dorsal iris margin is noticeable within a few days. Nuclei become rounded, and after about 5 days initiate DNA synthesis (Fig. 10-19). A lumen forms between the two lamina of the iris, and the cells begin to lose their pigment. At this stage, about 10 days after lentectomy, DNA synthesis and cell proliferation are in full swing. The ventral cell layer of the iris becomes multilayered, and in about 3 weeks, forms a body of tissue within which lens fibers differentiate. Synthesis of DNA declines, and the alpha, beta, and gamma crystallins are synthesized. Thus, during lens regeneration, differentiated iris cells transform into lens cells. To accomplish this transformation, the iris cells must synthesize DNA and divide. Figure 10-20 diagrams the events of this process. Here, again, regeneration is associated with a loss of the differentiated state (depigmentation) and with proliferation. Moreover, in lens regeneration metaplasia seems to occur, for the normal lens is formed from epidermis, whereas the regenerated lens is formed from iris, which itself is of neural origin.

## Mammalian liver regeneration

When a part of the liver of a mammal is removed surgically, rapid cell proliferation occurs in the remaining liver. These dividing cells make DNA, RNA, and also liver enzymes. Ultrastructurally, there is little evidence of dedifferentiation. However, there is excellent evidence to show that the regenerating cells pass through a new time sequence of gene function. The evidence for this is based upon nucleic acid hybridization. Labeled RNA precursor molecules were made available to regenerating livers, and at successive stages of regeneration, the newly formed RNA was analyzed by molecular hybridization. RNA

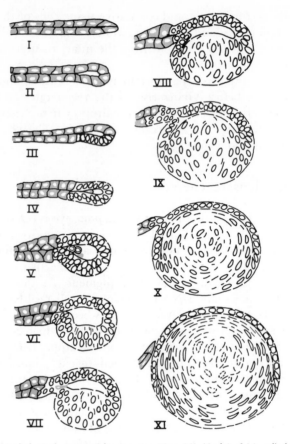

Fig. 10-19   Morphological stages of lens regeneration. I-II: Nuclei of iris cells become round; lumen forms. III-IV: Depigmentation of some iris cells, increasing number of free ribosomes in depigmented cells; V-VI: Depigmented cells divide rapidly and lead to a thickening of the new lens vesicle. VI-VIII: New fiber cell elongation begins in bottom portion of new lens; IX: New lens begins to be surrounded by flat epithelium which is formed through cell multiplication in the external cell layers; X-XI: Terminal differentiation of new lens, disappearance of iris stalk. (From Yamada. 1967. *In* Current topics in developmental biology 2:247–283. Fig. 1.)

from the same or from different stages of regeneration was used in competitive hybridization experiments. As the family of curves in Fig. 10-21 shows, the cells of the regenerate pass through a sequence of stages, each of which synthesizes different arrays of RNA; the final stage synthesizes RNA characteristic of the adult liver. Significantly, a similar sequence of RNA synthetic events occurs during fetal development of the liver; this was found when the RNA synthetic pat-

| Phase | Normal iris cell | Latent | Depig- mentation | Multiplication | Fiber differentiation | |
|---|---|---|---|---|---|---|
| Pigment granules | Abundant | Abundant | Decreasing | Almost absent | Absent | — |
| Mitosis | None | None | Initiated | Active | None | — |
| Nucleus | Small, irregular | Enlarging | Larger, spherical | Larger, spherical | Larger, ellipsoid | Disappears later |
| Nucleolus | Small, without granules | Small, without granules | Larger, with granules | Larger, with granules | Decrease in size and granules | Disappears later |

Fig. 10-20   Flow sheet of events in iris dedifferentiation and lens regeneration. (From Yamada. Table 1.)

Fig. 10-21   RNA synthesis during liver regeneration as measured by molecular hybridization. (a) Radioactive RNA from sham-operated animals was allowed to hybridize to DNA in the presence of increasing amounts of non-radioactive RNA prepared from sham-operated livers (●———●), 1 hour post-operative liver (○———○), 3 hour post-operative liver (□———□), 6 hour post-operative liver (▲———▲). Notice that all these RNAs are equally effective competitors. Thus, the RNA molecules present in normal liver are also present in regenerating liver. (b) Radioactive RNA from 1 hour post-operative livers was allowed to hybridize to DNA in the presence of increasing amounts of non-radioactive RNA prepared from sham-operated livers (●———●). Notice the low degree of competition in this experiment. If RNA from 1 hour post-operative livers is used, competition is excellent (○———○). RNAs from 12 hour post-operative (■———■) or 48 hour post-operative liver (▲———▲) gives intermediate values of competition. Thus, regenerating liver contains RNA that is not present in normal liver, and these RNA molecules characteristic of regenerating liver differ according to the stage of regeneration (From R. B. Church and B. J. McCarthy. 1967. J. Mol. Biol. **23**:*459–475.* Figs. 7b, 8a.)

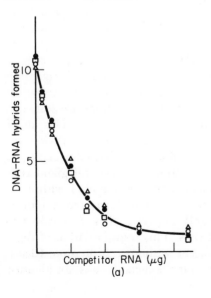

(a)

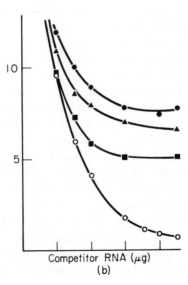

(b)

terns were studied during normal liver ontogeny, again using the technique of nucleic acid hybridization. Clearly, regeneration of the liver proceeds through a programmed sequence of gene activities similar to that of normal liver ontogeny. Liver regeneration requires extensive proliferation but not much loss of differentiation. As might be expected, metaplasia does not occur.

We might ask at this point whether cell proliferation is an absolute requirement for the cellular changes occurring during regeneration. Are there examples of differentiated cells losing their differentiated state and redifferentiating into new cell types without cell division?

## The lability of the differentiated state in Hydra

Hydra is a simple coelenterate that presents a number of advantages for developmental studies. For example, it contains only about 10 different types of cells. In this organism, it is therefore easier to trace the fate of a particular cell during regeneration than in other organisms where hundreds of cells are involved in regeneration. Some cells of hydra have remarkable totipotency, notably the interstitial cells, which can give rise to virtually all cell types of the mature organism. Even fully differentiated cell types of hydra, when isolated, have the capacity to regenerate normal adults. The gastrodermis, as analyzed histologically, is seen to consist of only two cell types: digestive and gland cells. When these two cell types are cultured together in a series of appropriate culture media, they can form an entire hydra. Metaplasia clearly occurs in this case, and it is important to know whether the metaplasia required cell division. It is possible to follow the developmental path of each of these two cell types. The gland cells, as shown by electron microscope analysis, lose their endoplasmic reticulum and shed their lipid inclusions. These structures characterize the differentiated state of this cell type. After these structures are lost, they can no longer be distinguished from interstitial cells. This dedifferentiation occurs without cell division. Next, the cells divide and their progeny differentiate into typical cnidoblast cells and nematocysts, highly specialized cells quite different from the ancestral gland cells.

When the digestive cells are isolated, the algal bodies within the cells break down. Next, muscle fibers begin to form and a mucous border characteristic of epidermal cells appears. These new cells now closely resemble normal epidermal cells. This transformation of digestive cells into epidermal cells can occur without cell division. Clearly metaplasia occurs in both cell types, but the metaplasia is not identical in both. The dedifferentiation of gland cells to form interstitial cells leads through cell division to the formation of very different cell types. In the transformation of the digestive cells into epidermal cells, on the other hand, no cell division is required. Rather, redifferentiation occurs directly. To our knowledge, this is the only example of direct metaplasia in the absence of cell division in a higher organism.

## Lethal cell differentiation

One of the most remarkable types of cell differentiation leads to cell death—that is, the terminal developmental stage is lethal. Such cells are programmed to die and their death is a normal characteristic of development. Cell death occurs in many parts of an organism, and in the embryo it is an essential feature of the normal morphogenesis of many organs. The union or detachment of parts of organs and the formation of a lumen in the center of an otherwise solid structure all require the death and disintegration of strategically placed cells. For example, cell death is essential to the separation of the fingers and toes from one another, to the separation of the lip from the gums, to the opening of the eyelids, and to the formation of the central canals of many ducts and organs. The failure of cell death to occur on schedule results in congenital abnormalities. One example, webbed toes, has already been described. Another is the occasional persistence of a small tail in human beings.

In other locations, the survival of embryonic cells that are normally programmed to die may lead to the development of embryonic tumors. For example, the medulloblastomas derived from the rapidly growing external granular layer of the fetal cerebellum may represent surviving embryonic cells that should have died. Other tumors of embryonic origin that may develop in a similar fashion are the sympathetic neuroblastomas, retinoblastomas, nephroblastomas, hepatoblastomas, and embryonic sarcomas. These tumors are composed of aberrant cells that originated either from transformed normal cells or from embryonic cells that failed to die on schedule.

A conspicuous example of the role played by cell death in the normal development of organisms is the regression of larval organs during metamorphosis. The cells of the tail of the tunicate tadpole or the frog tadpole die and are resorbed as the larva transforms into an adult. Many other, less conspicuous, larval organs also regress during metamorphosis. Since all cells of the developing embryo have the same genetic makeup, the regression of certain cells and the persistence of others must flow from their state of differentiation on receiving an appropriate stimulus. A clear case involves the regression of cells in the Müllerian ducts of males during hormonal maturation. Mammalian embryos are equipped with the rudiments of the sexual organs of both sexes; one set is suppressed as a consequence of hormonal stimulation in accord with the genetic makeup of the individual. Likewise, during thyroxin-stimulated metamorphosis of frog larvae, some cells die and others proliferate. The response of the cell is thus highly specific to its state of differentiation.

In the adult, too, cell differentiation can lead to cell death. The erythroblast of adult bone marrow passes through a sequence of developmental steps that result in the loss of its nucleus; therefore, it soon dies. Likewise, the differentiation of the epidermis in vertebrates leads to a heavily keratinized cell that eventually dies and is sloughed off. Many secretory cells likewise die in the process of releasing their secretions. These are all normal types of cell differentiation that terminate in death. Similar sequences of differentiation can be induced experimentally. Removal of embryonic limbs, for example, will bring about the death of many cells in the central nervous system that are normally responsible for innervating those limbs. Thus, these nerve cells require a highly specific functional relationship with the innervated cells to survive. Simple nutrition can scarcely be involved. The functional dependence can best be understood as that of maintaining the nerve cell in a stable equilibrium state of differentiation. Once this equilibrium is upset, the nerve cell begins to change and enters a lethal pathway of differentiation.

Like all steps in cellular differentiation, the time at which the differentiated step leading to cell death is taken occurs at a specific stage in the life of the cell. This has been clearly demonstrated in experiments on the wing bud of developing chick embryos. The cells attaching the posterior margin of the wing bud to the body in the young chick embryo are normally destined to die as the wing bud separates from the body. These cells can be removed from the area in which they normally reside at an early stage of development and transplanted to the dorsal surface of the wing bud. In this new location, they do not die but become an integral part of the dorsal wing surface. If, however, these cells are allowed to remain in their original position beyond a critical developmental stage (stage 22), then when transplanted to a new site they nevertheless continue their differentiation to death. Thus, the initial steps of differentiation are not irreversible, but at some point along the pathway the lethal direction of differentiation becomes fixed, at least under all experimental conditions tested so far.

Like all other cell processes, the steps in cell differentiation leading to cell death are subject to gene control and can be changed by mutation. For example, webbed toes in human beings is due to a mutation that permits survival of the cells between the toes, even though genetically normal cells in this location would die. There are also many examples of mutations that induce cells in particular areas to die at specific stages in development; cells with normal genotypes would have remained viable in those locations. For example, human beings with XO and XXY genotypes differentiate normal cells in most organs of the body, but viable gametes do not differentiate in the gonads. The developing germ cells die at an advanced stage of differentiation but other cells of the body develop essentially normally. In the XXY individuals, the germ cells die in early childhood or at an early stage in puberty. The primordial germ cells with these abnormal karyotypes are incapable of completing their differentiation, although early steps in the pathway of gamete differentiation remain open to them. More severe karyotypic abnormalities—for example, trisomy for chromosome two—permits the embryonic cells to take part in the formation of an apparently normal placenta and certain fetal membranes, but these cells cannot form the body of an embryo.

Single gene changes controlling cell death are also known. The tailless mutant of the mouse is a clear example. In such mice, a constriction appears in the proximal part of the tail at about 11 days of embryonic development, and a localized zone of cell death occurs. The distal part of the tail then degenerates, and the mouse is born with only a short stump. Rumplessness in fowl embryos exhibits a similar form of degeneration. In these embryos, the presumptive tail tissue degenerates even before the onset of tail formation, and the newly hatched chick is completely without a tail.

The gene for retinal degeneration in mice affects only certain specifically differentiated cells, the rod cells of the retina. These cells begin to differentiate, but before differentiation has been completed, the distal parts of the cells cease development and degenerative changes begin. It is important to note that these cells develop normally in the mouse throughout embryonic life and for the first 14 days after birth. No other type of cell appears to be affected by this gene.

Several different genes, at least six are known, bring about the degeneration of the cells in the labyrinth of the ear, inducing a kind of waltzer-shaker syndrome. The labyrinth of such mice is normal at birth, but abnormalities appear as histological differentiation approaches maturity about 12 days after birth. At this time, the organ of Corti, the spiral ganglion, and other associated tissues degenerate. The precise pattern of degeneration varies slightly in accord with the specific gene that has mutated. These inherited patterns of cell death illustrate very well the general principle that only a portion of the cell's genome is brought into play at any one time. If the part called upon does not contain a deleterious mutant gene, then the cell survives and continues to differentiate in a perfectly normal fashion. If, however, a mutant gene is called upon to function at some stage in the differentiation of the cell, then the mutant effect will become apparent. If the normal role of the gene is indispensable to the survival of the cell under those conditions of differentiation, then the cell with the mutant active gene will, of course, degenerate and die. It is interesting that such a cell does not simply enter an arrested state of differentiation, as might be expected when a mutant gene fails to perform normally. Instead, the general metabolic organization of such differentiating cells seems to have reached a state in which the normal function of the mutant gene is indispensable for survival and not simply for the next step in differentiation.

REFERENCES

Burnett, A. L. 1968. The acquisition, maintenance, and stability of the differentiated state in hydra. *In* H. Ursprung [ed.] The stability of the differentiated state. Springer, New York.

Church, R. R. and B. J. McCarthy. 1967. Ribonucleic acid synthesis in regenerating and embryonic liver. J. Mol. Biol. 23:459–475.

Gehring, W. 1968. The stability of the determined state in cultures of imaginal disks in *Drosophila*. *In* H. Ursprung [ed.] The stability of the differentiated state. Springer, New York.

Grueneberg, H. 1952. The genetics of the mouse. Nijhoff, The Hague.

Hadorn, E. 1961. Developmental genetics and lethal factors. John Wiley & Sons, Inc., New York.

Hadorn, E. 1967. Dynamics of determination. *In* M. Locke [ed.] Major problems in developmental biology, pp. 85–104. Academic Press, New York.

Hay, E. D. 1968. Dedifferentiation and metaplasia in vertebrate and invertebrate regeneration. *In* H. Ursprung [ed.] The stability of the differentiated state. Springer, New York.

Landauer, W. 1954. Chemical production of developmental abnormalities and of phenocopies in chicken embryos. J. Cellular Comp. Physiol. 43:261–305.

Lewis, E. B. 1964. Genetic control and regulation of developmental pathways. *In* M. Locke [ed.] The role of chromosomes in development, pp. 231–252. Academic Press, New York.

Ursprung, H. 1963. Development and genetics of patterns. Am. Zool. 3:71–86.

Yamada, T. 1967. Cellular and subcellular events in Wolffian lens regeneration. *In* A. A. Moscona and A. Monroy [eds.] Current topics in developmental biology, vol. II, pp. 247–283. Academic Press, New York.

# ELEVEN

## Abnormal Development
## and Cancer

### Neoplasia

It is not surprising that embryonic development, including cell differentiation, has its pathological manifestations. All normal biological processes occasionally go awry. Neoplastic cells, although abnormal, are simply different types of differentiated cells. Neoplasms are commonly classified as benign or malignant. This classification tends to obscure the more fundamental properties of tumor cells. In an attempt to understand the basic properties of tumor cells, let us examine three characteristics of cells: their rate of division; the adhesive properties of the cell membrane, a characteristic that probably determines whether a cell remains in a tissue or moves about in the body and invades other tissues (metastasis); and cellular metabolism. These three characteristics can vary quantitatively over a large range in different cell types and furthermore exist in different combinations, both in neoplastic and normal cells. It is important to realize that probably no gene codes for cancer per se. Rather, gene activity in the tumor cell seems to be misprogrammed by epigenetic or genetic mechanisms to produce the tumorous state of differentiation.

#### Levels of misregulation

In previous sections, we have discussed the various levels at which gene action is regulated. The properties of neoplasms illustrate misregulation at one or all of these levels. In those cases in which mutant genes are clearly responsible for converting normal into neoplastic cells, such as in retinoblastomas, these mutant genes probably act on regulatory mechanisms, altering the normal temporal and spatial programming of the genome.

An essential characteristic of neoplasms is continued cell division beyond the time at which it comes to a halt in the normal case. Excessive rates of cell division have been described in many neoplasms, but normal cells really exhibit the same range of division rates. In fact, many normal cells, such as blood-forming cells, divide more rapidly than some malignant cells. Abnormally persistent cell division by itself produces a benign neoplasm, although such a neoplasm can become fatal by continued growth. For example, there are cerebral gliomas that look histologically benign but inflict lethal mechanical damage on the brain. In so-called gliosis, which often results from cerebral injury, the same cells proliferate as in gliomas. Proliferation comes to a halt, whereas in gliomas it continues. A normal cerebral scar thus differs from a tumor because in the latter, cell division persists. Of course, a histologically benign neoplasm need not be fatal. For example, leiomyomas can grow indefinitely without producing deleterious effects other than the discomfort of their mass.

The second important aspect of neoplasia, cell migration, is observed in many normal embryonic cells and even in some normal adult cells. One of the best-studied cases is that of melanoblasts. These cells originate in the neural crest of the embryo and then migrate through several tissues before reaching their terminal locations, where they complete their differentiation and become nondividing, nonmigrating melanocytes. This migratory behavior is a prerequisite for normal distribution of pigment cells and in no way has deleterious effects, because the melanoblasts, after reaching their final location, do not continue to divide. Abnormal paths of migration may produce abnormally located pigment spots, such as brown nevi. Only when division and migration are both persistent does a melanoma, one of the most malignant cancers, result. Thus, a perfectly harmless smattering of pigment spots, or a malignant melanoma, may form, depending entirely upon whether or not these cells continue to divide.

The third parameter—namely, the biochemical activity of the cell— is of fundamental importance for understanding neoplasms as expressions of the misprogramming of gene function. Consider, for example, the phaeochromocytomas. These are tumors of the adrenal or extra-adrenal chromaffin tissue. They can be fatal for the host organism when they are still very small, even microscopic, because of excessive secretion of epinephrine and norepinephrine. Little or no cell migration resulting in metastases occurs in these tumors, and only minimal cell division is required to amplify the secretory activity of the neoplasm to harmful levels. Small lethal tumors of this type secrete more hormone per cell than do normal adrenal cells, and large tumors must secrete substantially less per cell than normal tissue. Thus, the level of secretory activity need not be exorbitant or erratic on a per cell basis, just abnormally large for the entire tumor.

*Misregulation at more than one level; and the degree of misprogramming*

Although it is true that a minimal rate of cell division is necessary for a tumor to be recognized, neoplastic cells probably should be characterized principally by their abnormal migratory or biochemical behavior. These characteristics vary greatly in different neoplasms. Migratory behavior varies from nearly zero to that maximally observed in normal adult cells. Biochemical activity ranges from benign to virulent. But, in essence, variations in any of the three fundamental characteristics can be attributed to misprogramming of gene function.

A combination of continued cell division and biochemical lesion is represented in parathyroid carcinomas. These affect the host by excessive proliferation in situ and by secretion of excessive amounts of parathyroid hormone.

A different combination of the three basic characteristics is found among the insulin-secreting islet cell carcinomas. These are characterized by migration (metastasis) and production of excessive amounts of insulin. Proliferation is minimal because the rate of cell division is low and the lethal effect takes place while the tumor is still small.

Abnormality of all three of the basic cell characteristics is observed in the case of hormone-secreting nonendocrine lung tumors. The cells of these tumors divide rapidly, metastasize, and secrete abnormal quantities of various hormones. Each characteristic by itself, of course, is a normal property of some normal cell. It is only in this abnormal combination that these normal cell properties produce neoplasms.

The degree of misprogramming can cover a wide range, which is well illustrated in the broad spectrum of conditions connected with excessive erythroid proliferation. Erythroleukemia, for example, stands at one end of the spectrum in which erythroid stem cells migrate extensively and proliferate rapidly and persistently. In polycythemia vera, on the other hand, there is persistent, excessive division of erythroid stem cells but migration is insignificant. At the normal end of this spectrum of erythrocyte behavior, is polycythemia induced by high altitude, which represents a normal response to physiological conditions simply requiring enhanced cell division.

It thus appears safe to say that the biological machinery of neoplastic cells does not differ in any qualitative way from that of normal cells. Rather, neoplasia should be viewed as a result of misprogramming of gene function leading to quantitative abnormalities. Thus, a continuous spectrum of migration and division in response to environmental influences occurs among cells, ranging from normal responses to physiological conditions to seriously abnormal behavior and malignancy.

## Teratomas

Teratomas are neoplasms that apparently arise from primordial germ cells. They are chaotic arrangements of many different cell types and tissues; frequently, the cells are organized as functioning tissues. The differentiated tissues derived from the teratoma stem cells usually proliferate at reduced rates, or not at all, and are not malignant. The stem cells themselves continue to proliferate and give rise to a variety of differentiated tissues as well as to successive generations of malignant stem cells.

A genetic basis for teratomas has been demonstrated in mice. In one strain of mice, testicular teratomas occur very frequently. Of course, not all mice in this strain develop teratomas and not all primordial germ cells in any one mouse become neoplastic. In fact, only an extremely small fraction are transformed. Some threshold environmental stimulus must be necessary to transform germ cells into the teratoma stem cells. Nevertheless, the genetic predisposition to malignant differentiation has been clearly established by breeding experiments. Embryonal carcinoma cells are the multipotential stem cells of teratocarcinomas. This was shown by single-cell cloning experiments from dissociated teratomas. Individual clones in this experiment varied in their developmental potentiality, but all gave rise to a variety of differentiated tissues.

Transplanted teratomas in mice of this strain occasionally develop into large cystic embryonal bodies devoid of embryonal carcinoma cells. These embryonal bodies differentiate into many kinds of tissue: cartilage, bone, muscle, nervous tissue, and various glands of endodermal origin. However, these embryonal bodies stop growing when their constituent cells differentiate. Evidently these cells are not malignant, even though their progenitors were.

Of all the tumors analyzed, teratomas represent a unique kind of aberration in gene programming. Not only are the specific gene programs out of order in teratomas, but the basic mechanism of programming itself appears defective. The "neoplastic event" in teratoma formation must occur during the differentiation of the primordial germ cell. The occurrence of a neoplastic event at this early stage suggests the existence of a single basic common mechanism for programming genes. A defect in this postulated mechanism would produce the chaotic arrangement of tissues seen in teratomas but would not prevent normal differentiation of specific cell types once the initial defect had been passed. That is, specific programs for regulating gene function in highly differentiated cells can still develop in teratomas, but the selection of a particular program appears uncontrolled and erratic. What this might mean in molecular terms is completely unknown. Perhaps a closer look at teratoma formation would provide useful clues in searching for biochemical mechanisms involved in the differentiation of normal cells.

## Carcinogenic agents

To be consistent with our view that neoplasia is a result of misprogramming of genes, carcinogenic agents must ultimately affect gene function. But at what level and by means of what molecular events? There are at least four recognized causes of cancer that should be considered: gene mutation, physiological stress, treatment with a carcinogen, and viral infection.

### Genes and cancer

Dominant autosomal mutations are responsible, for example, for retinoblastoma in human beings. These genes, although present in all the cells of an affected individual, manifest themselves only in retinal cells and only in a few of them. Retinal cells are not vulnerable until they have reached a relatively advanced stage of differentiation.

A polygenic basis for neoplasia has been demonstrated in some tumors of inbred strains of mice. These tumors are based on the presence of several genes in specific allelic form. Individually, these genes are harmless, as demonstrated by outcrossing the susceptible strains to unrelated mice. But the collective effect of the several genes shifts the metabolic balance toward neoplastic differentiation. This shift apparently occurs in only a very few cells, presumably those that have reached highly specific states of differentiation.

Unusual combinations of genes can be obtained by hybridizing distinct species. Hybrids between platyfish and swordtails (*Xiphophorus maculatus* × *Xiphophorus montezumae*), for example, almost invariably develop melanomas, derived from the macromelanocytes. All other cell types, including micromelanocytes and other pigment cells, remain essentially normal in these hybrids. Apparently, the interaction of several genes rather than the effect of any single gene is the decisive requirement for malignant transformation in cells that have reached a specific stage of differentiation—that is, have become macromelanocytes. No other cell type becomes neoplastic, even though all the cells of these hybrid fish have the same genetic makeup.

The melanomas of hybrid fish, and also the strain-specific tumors of inbred mice, are not the result of new genes or novel reactions but of specific combinations of genes. The genetic basis for neoplasia does not depend upon unique cancer-producing genes, but rather upon the abnormal programming of normal gene function.

### Physiological stress

Transplantation of the ovary into the spleen eliminates the feedback control regulating hormonal synthesis in the ovary and produces an intense pituitary stimulus to the ovary. Not infrequently, the consequence of this stimulus is neoplasia of the ovary. Physiological stress simply appears to force the cell's metabolism into abnormal patterns through stimulation from outside. The resulting abnormal patterns of metabolism can produce a lasting transformation of the cell's gene function, so that the abnormal metabolism persists when the initiating stimulus is removed.

### Carcinogens

Carcinogens, like the molecules involved in physiological stress, change the metabolic activities of the recipient cells. The specific cell characteristics induced by carcinogens probably arise during normal cell differentiation too, but not in the same sequence or combination. This is particularly true for adhesivity and rate of cell division. The precise mode of action of carcinogens is presently unknown, although much research has been directed to this problem. Two obvious areas of possible involvement are the genes and the surface membrane of the cell. If a carcinogen altered the cell membrane so that it lost its adhesivity, then malignant transformation would result in all normal cells that continued to divide.

## Oncogenic viruses

Viral infection changes the metabolic activity of cells. Accordingly, the oncogenic effect of viruses can easily be fitted into the preceding explanations based on the assumption that changes in metabolic activity at critical periods can lead to neoplastic differentiation. Viruses may function as simple provoking stimuli to transformation and then be lost from the transformed cell, or the continued presence of the virus may, in some instances, be required to provide the cell with a constant supply of new molecules coded by the viral genome. These viral molecules could maintain the neoplastic pattern of metabolism.

## Neoplasia as a disease of cell membranes

The adhesive characteristics of cells are decisive in regulating cell association. Whether cells take up special positions with reference to adjacent cells, or remain fixed in position, or continue to migrate is of critical importance in embryonic development. Likewise, the capacity of malignant neoplasms to metastasize surely depends upon the adhesive properties of the cell membrane. Morphogenesis in normal development requires extensive shifts in the relative positions of many cells. Accordingly, cell membranes must undergo changes in adhesive properties. This can be demonstrated by dissociating embryonic tissues, mixing the cells in various combinations, and then examining the behavior of the different kinds of cells in the mixed aggregates. The cells, originally randomly dispersed through the aggregate, sort themselves out through migration and selective adhesion. Liver cells preferentially adhere to liver, cartilage to cartilage, kidney to kidney, and so forth. Moreover, the relative positions of kidney, liver, and cartilage in the aggregate are also specified by the properties of the individual cells. Since the cells are motile and in contact with one another, the most adhesive cell type inevitably moves to the center of such mixed aggregates, and progressively less adhesive cells are pushed toward the periphery. This behavior of embryonic cells emphasizes the fact that the cell membrane can be structured to express many different degrees of adhesiveness and thus can regulate the specific associations of cells through encouraging or preventing migration.

In the transformation of normal cells into metastasizing neoplasms, the cell membrane becomes less adhesive, enabling the neoplastic cell to dissociate from the surrounding cells and to infiltrate other organs.

It is precisely this infiltration, coupled with continued proliferation, that makes such neoplasms malignant. However, the surface membrane of such cancer cells is probably very similar to the cell membranes of many kinds of embryonic cells. At least one kind of normal adult cell also has a nonadhesive surface membrane that permits it to migrate extensively through other tissues of the body. This is the leukocyte. If the normal migration of leukocytes were coupled with cell division in the peripheral locations to which they migrate, they would certainly prove to be malignant, but fortunately they do not divide. Thus, it is clear that cancer cell membranes, although relatively nonadhesive, are not necessarily abnormal. The malignant abnormality lies merely in coupling the production of nonadhesive membranes to continued cell division.

How do membranes grow and acquire their specificity? Membranes must grow as the cell grows. By the time the cell divides, the total amount of cell membrane has essentially doubled. In addition to growing, the cell membrane also changes in specific structure as cells differentiate. Two sources of such changes in structure should be considered. First, if cell membranes of different cells are characterized in part by the presence of different kinds of molecules, then differential gene function with reference to membrane molecules would be essential to the differentiation of new membranes. Second, changes in membrane structure may be achieved simply by changing the proportions of membrane molecules or their topographic relationship to one another. In this case, new gene activity would not be necessary to obtain changes in membrane properties.

Very little information relevant to this issue has been obtained so far for metazoan cells, whereas extensive research on the pellicle of protozoan ciliates has dramatically established a remarkable degree of autonomy of the pellicle (perhaps homologous to cell membranes) from the genome (see p. 204). Whether or not metazoan cell membranes are similarly autonomous in their mode of reproduction is a question well worth keeping in mind. If the structure of metazoan cell membranes is determined by the template activity of preexisting membranes, then some accidental variation in membrane structure might lead to a heritable loss in adhesivity. This possibility can be tested today through the procedures of cell fusion in tissue culture, in which it should be possible to combine the cell membrane of one type of cell with the genome of a very different kind of cell. In fact, malignant and normal cells have already been fused to form a single hybrid cell. The fused cell was neoplastic. However, this elegant experiment did not resolve the issue of membrane autonomy because the fused cell contained the genomes of both the malignant and the normal cell. The properties of the membrane in the cell may have been only a reflection of the dominant activity of the genome from the malignant cell in structuring the membrane rather than an expression of membrane autonomy. To resolve this issue, we must produce a cell equipped with the nucleus of a normal cell and the membrane of a malignant cell. We have every reason to believe that this experiment is feasible and will be carried out in the near future.

## REFERENCES

Braun, Armin C. 1969. Understanding the cancer problem. Columbia University Press, New York.

Finch, B. W. and B. Ephrussi. 1967. Retention of multiple developmental potentialities by cells of a mouse testicular teratocarcinoma during prolonged culture *in vitro* and their extinction upon hybridization with cells of permanent lines. Proc. Nat. Acad. Sci. 57:615–621.

Markert, C. L. 1968. Neoplasia: A disease of cell differentiation. Cancer Res. 28:1908–1914.

Mintz, B. and G. Slemmer. 1969. Gene control of neoplasia. I. Genotypic mosaicism in normal and preoplastic mammary glands of allophenic mice. J. Nat. Cancer Inst. 43:87–95.

Pierce, G. B. 1967. Teratocarcinoma: Model for a developmental concept of cancer. *In* A. Monroy and A. A. Moscona [eds.] Current topics in developmental biology, vol. II. Academic Press, New York.

Pitot, H. C. 1966. Some biochemical aspects of malignancy. Ann. Rev. of Biochem. 35:335–368.

Stevens, L. C. 1967. The biology of teratoma. *In* M. Abercrombie and J. Brachet [ed.] Advances in morphogenesis, vol. VI. Academic Press, New York.

Stevens, L. C. 1967. Origin of testicular teratomas from primordial germ cells in mice. J. Nat. Cancer Inst. 38:549–552.

Symposium on the developmental biology of neoplasia. 1968. Cancer Res. 28:1797–1914.

# TWELVE

## Omnis Embryo e DNA?

The dominating role of genes in embryonic development has been made abundantly clear in preceding chapters of this book. One might reasonably conclude that all of the characteristics of the developing embryo and of the adult were completely encoded in the sequence of nucleotides making up the DNA of the genes. However, there is reason to believe that some of the information required to make an organism is not encoded in DNA. Information may also be stored in metabolic systems or in macromolecular aggregates. The most dramatic evidence for such possibilities comes from research on the protozoan ciliates, *Paramecium* and *Tetrahymena*. These single-celled ciliates undergo both vegetative and sexual reproduction. During vegetative reproduction, the protozoan divides into two daughter cells that are indistinguishable from each other and from the preceding parental cell. Such vegetative reproduction can continue for many generations to produce thousands of individuals all descended from a single individual that may have been the product of sexual reproduction.

During sexual reproduction, two different ciliates come together in conjugation and exchange micronuclei through an aperture that forms between the conjoined surfaces. The exchanged micronuclei fuse with a second haploid nucleus present in each individual, thus reestablishing the original diploid chromosomal complement of the nucleus. On separating from one another, the two individuals participating in sexual reproduction are genetically identical. Occasionally during this process of separation, an accident occurs and a portion of one protozoan is carried away by the separating partner. A dramatic instance of this kind of accident occurs when one individual tears away the mouth

203

parts of the partner. Such an individual thereafter is equipped with two mouths (Fig. 12-1). When vegetative reproduction occurs in such a two-mouthed individual, the descendant offspring have two mouths, and this abnormal characteristic is perpetuated through many generations. Such individuals can be mated with normal protozoans; when the conjugating individuals separate, each is genetically identical, but one of the exconjugants has two mouths and the other only one. Thus, it is clear that the two-mouthed condition is not dependent upon nuclear genes.

The mouth region is largely a structure of the pellicle, and in many respects the pellicle seems to be a self-perpetuating structure. Many pellicular aberrations are passed on from generation to generation during vegetative reproduction (Fig. 12-2). Thus, the organization of the pellicle appears to be substantially independent of the genetic makeup of the nucleus. Its structure reflects in part the accumulated accidents that occurred in preceding generations. Of course, any induced change

Fig. 12-1   Normal Paramecium showing one mouth (a), and abnormal vegetatively reproducing individual with two mouths (b). (Courtesy of T. M. Sonneborn.)

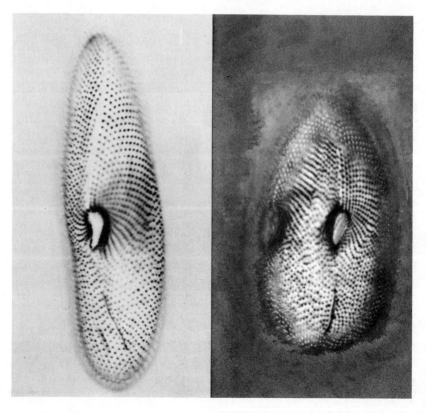

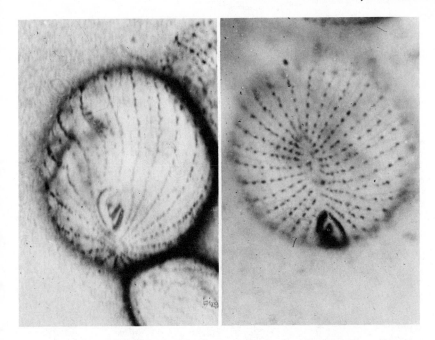

Fig.12-2  By accidents of development the number of ciliary rows in tetrahymena may be changed. Once an alteration has occurred, however, it persists through many generations of cell division. On the left is an oblique anterior view of an individual with about 18 ciliary rows; on the right is a polar view of an individual with about 30 ciliary rows. These different numbers persist through cell division independently of the chromosomal genotype. (Courtesy of D. L. Nanney.)

in pellicular organization must be compatible with survival and must be structurally reproducible if it is to persist. The pellicle evidently acts as a self-perpetuating template for the assembly of molecules, many of which are formed under the aegis of the nuclear genes, into structures identical to the preceding templates.

Two different kinds of historical accidents may, therefore, become embedded in the structure of an organism and bring about its gradual evolution. One kind of accident involves the substitution of one nucleotide for another in the DNA, and the second kind of accident involves the assembly of molecules into novel structures. These molecular arrangements, although obviously consistent with the intrinsic properties of the molecules, might not occur with sufficient frequency to have any significance for the life of a single individual. Only rarely would they occur, but if they served as self-replicating templates, then they would be passed on to succeeding generations. The pellicle of ciliates provides a clear example of such vegetative inheritance, and perhaps a similar mode of inheritance applies to other cell organelles, such as mitochondria, chloroplasts, and possibly the cell membranes.

Whether this kind of inheritance occurs in the embryonic development of metazoans is unknown. However, there is at least one instructive case that relates to the same general problem. It is the coiling of
the shell in the snail *Limnaea.* These snails exist in two genetic varieties,
one coiling to the left, the other to the right. Breeding experiments
have shown that the right-coiling form is dominant. What is remarkable about these snails, however, is the fact that the coiling of any
particular individual is not a function of the genotype of that individual, but is in accord with the genotype of the maternal parent.
The inheritance pattern in these snails is shown in Fig. 12.3. The fact
that an individual genetically homozygous for left-coiling can still
coil to the right is a dramatic example of what has been called maternal
inheritance. Apparently, the organization of the egg in the maternal

Fig. 12-3   Inheritance pattern of coiling in *Limnaea.*

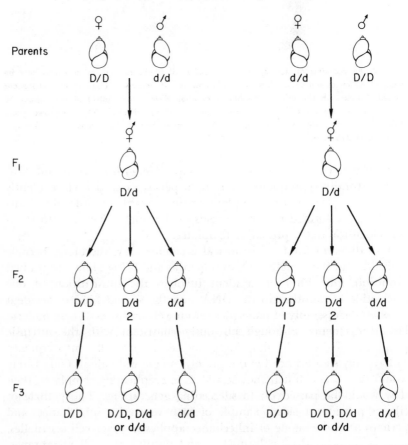

ovary achieved under the direction of the genes for right-coiling persists in the subsequent division of the egg and in the formation of the embryo, without any regard for the genetic makeup of the embryo. A cytoplasmic characteristic can thus be imposed upon the egg and can persist and be replicated in the absence of those genes initially responsible for the characteristic. The genetic origin of the characteristic seems less important than the maintenance and transmission of the characteristic through many cell generations in the absence of immediate genetic control.

The same general philosophical problem in different form can be illustrated on the molecular level by the behavior of the subunits of lactate dehydrogenase in certain fish. The evidence as presented earlier indicates that in nearly all cases, the subunits of LDH assemble at random within the cell in accord with their own characteristics. Such assembly fits well with the notion that the structure of the tetrameric LDH molecule is, in fact, a direct reflection of the genes responsible for encoding the LDH subunits. However, there is at least one group of fish, represented by the alewife, in which the assembly of LDH subunits does not appear to follow a binomial distribution. This pattern is shown in Fig. 12-4. Obviously, LDH-2 and LDH-4 are present in insufficient amounts to fit the requirements of a binomial distribu-

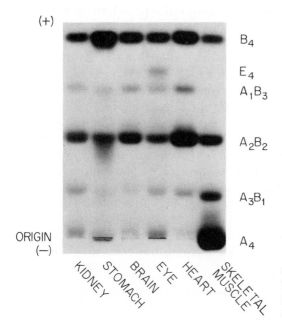

Fig. 12-4 Zymogram showing LDH isozyme banding pattern in Alewife (*Alosa pseudoharengus*).

tion. Yet in vitro dissociation and recombination of these isozymes does yield a binomial distribution. Thus, the physicochemical characteristics of the subunits do enable them to associate at random to form the expected tetramers. The fact that they do not do so in vivo indicates that the cellular organization restricts the combination of the subunits so as to produce the nonbinomial distribution of isozymes. Whatever this mechanism may be, it specifies the organization of individual gene products into structures of a higher order.

Neither the nonbinomial assemblage of LDH subunits nor the coiling in snails constitutes a clear-cut example of self-replicating templates analogous to the pellicle of protozoans. However, the existence of these various characteristics strongly suggests caution in assigning to the DNA of the genes total responsibility for the organization of the egg and for its subsequent development into an adult individual.

## The genetic control of order and shape

The various types of cells in a higher organism are obviously not arranged randomly. Rather, cells are organized into tissues and organs of well-defined size and shape. A number of embryological mechanisms are known to account, at least in part, for this observed order. Embryonic induction is one such mechanism. Dependence upon inductive stimuli tends to assure a harmonious integration of diverse developmental pathways, so that each cell type is formed at the right time and in the right place. Morphogenetic movements constitute another mechanism, by which order and shape are brought about during development. Both of these mechanisms, which are discussed in more detail in another volume of this series,* can clearly be classified as "epigenetic." The former brings about specific types of differentiation by extrinsic, diffusible substances; the latter relocates cells of different developmental fates with respect to each other and thus indirectly permits sequential transient influences to become effective. From this discussion, it is clear that genes may influence shape and order in the organism by interfering with morphogenetic movements or by the release of inductive stimuli. Indeed, mutations are known that affect development in just this way. Teratological development, such as spina bifida or cleft palate in humans, results from abnormal morphogenetic movements. We have already mentioned malformation or, in extreme cases, absence of the kidneys resulting from the failure of an

* J. P. Trinkaus, *Cells Into Organs* (Englewood Cliffs, N.J.: Prentice-Hall, Inc.), 1969.

inductive stimulus in the mutant mouse known as Danforth's short tail.

There is a quite different way in which observed nonrandom arrangements of different cell types can be brought about. Specific types of cell differentiation may be initiated locally and autonomously in precursor cells without any influence from other cells. A plant leaf provides an obvious example. Its surface is essentially an epithelium composed of one cell type. Arranged on this surface are the stomata of the leaf, simple breathing apparatuses that consist of cells different from those forming the epithelium surrounding them. There are, for example, contractile cells that permit the stomata to open and close. This type of differentiation—that is, into stomate cells—must have been initiated locally in stomate-progenitor cells. Stomate cells cannot have been formed elsewhere in the plant and then moved to their final, ordered positions in the leaf, because cell migration does not occur in developing leaves. Thus, stomate formation appears to be initiated locally, through as yet unknown mechanisms.

Are genes known to determine the pattern of such locally initiated differentiation? We have discussed numerous examples of genes affecting the quality of differentiation per se. What we are asking now is whether genes also control where in the organism a given type of differentiation is initiated.

Materials particularly well suited for studying this question are found in the bristle patterns of the fruitfly *Drosophila.* You will remember from earlier sections in this book that the integument of adult flies is derived, during metamorphosis, from imaginal discs. Histologically, these discs are flat epithelia consisting of one or, at most, very few cell layers. All the cells of the epithelium have the same ultrastructure and apparently are not differentiated in the larval stage. Why is it, then, that only some of these cells give rise to bristles, and always in the same geometrical pattern? A glance at Fig. 12-5 shows immediately that the bristles on a *Drosophila* thorax, for example, occur in sharply defined areas. Do the progenitor cells that lead to the formation of these bristles migrate into the appropriate positions and then differentiate? Or is "bristleness" initiated locally, without cell migration, and in a reproducible pattern? No conclusive answers are available at present, but it is interesting that Mendelian genes that influence the bristle pattern are known. One of these "pattern mutants" has been named "achaete," since it lacks the so-called posterior dorsocentral bristle. Other features of this mutant are that some bristles are arranged in a pattern different from wild type (Fig. 12-5).

The goal underlying the study of the genetic control of this altered

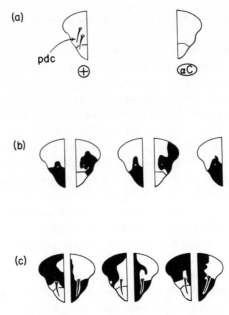

Fig. 12-5 Diagrams of half-thoraces of *Drosophila* resulting from somatic crossing over. Note the location of the posterior dorsocentral bristle (pdc, arrow). For the mosaic experiments, conditions were chosen that lead to the absence of pdc (right). (b and c) Mosaic half thoraces consisting of wild-type (white) and achaete (black) tissue.

pattern is to examine tissues that contain both mutant and wild-type cells. Such mosaic tissues cannot be produced by surgical grafting methods. It is possible, however, to obtain such mosaic tissues by genetic methods similar to those described earlier for the *Drosophila* eye (p. 74)—that is, through the induction, at an early stage of development, of somatic crossing-over. Figure 12-5 shows diagrams of half-thoraces of mosaic flies resulting from such experiments. Clearly, a posterior dorsocentral bristle appears only if the appropriate region of the thorax is occupied by wild-type tissue and fails to appear if this region is occupied by mutant tissue. Thus, the gene responsible for the achaete trait works very topically itself and certainly does not control pattern in a systemic sense.

## Molecular self-assembly

The title of Chapter 12 asks whether the whole embryo, not only its constituent macromolecules, is ultimately encoded entirely in the DNA contained in its cells. When we examine order and shape in the embryo, we are persuaded to answer the question in the affirmative for cases such as the bristle pattern in *Drosophila,* but for single-celled organisms such as *Paramecium,* shape and cortical organization appear to be propagated without the intervention of DNA; alternative arrangements are possible with exactly the same DNA.

In a general sense this conclusion is true for epigenetic modifications of gene products. The well-studied enzyme, ribonuclease (RNase), provides a good example. The primary structure of this enzyme is known. Its tertiary structure is stabilized by four disulfide bridges. When these bridges are destroyed by reduction with urea, the molecule loses its tertiary structure and also its biochemical activity. Conditions have been found under which such reduced RNase can be reoxidized to a biochemically active enzyme, and physical-chemical evidence indicates that it has regained its normal tertiary structure during this oxidation. If, on the other hand, these conditions are not met precisely so that sulfhydril pairing does not occur at the right places, then biochemically inactive molecules form. Obviously, DNA information does not intervene directly in regulating the correct versus incorrect renaturation. All that DNA contributes to the RNase is the specification of the primary structure of the protein, thereby limiting the range within which higher orders of structure may be formed. Precisely which of the higher orders of structure is formed is dependent on other environmental parameters such as pH and salt concentration.

Ribonuclease is by no means the only gene product known to assume various configurations which depend on the ionic environment. Similar denaturation-renaturation experiments have been conducted successfully on even more complex biological structures, such as ribosomes and bacterial flagella. *Escherichia coli* ribosomes, for example, can be disrupted into a series of smaller, biologically non-functional components. The resulting particles can then be renatured in vitro to form active ribosomes again. Bacterial flagella can be disaggregated into proteins called flagellin with a molecular weight of about 40,000. At the right pH and salt concentration, flagellin spontaneously reaggregates to form flagella that are indistinguishable under the electron microscope from the original flagella.

Thus, macromolecules can assume a number of different conformations limited by their primary structure which was specified by DNA. Which of these conformations is assumed, however, is determined by the milieu in which the molecule resides. In this sense, a large portion of development on each level of biological organization is clearly epigenetic.

## REFERENCES

Abram, D. and H. Koffler. 1964. In vitro formation of flagella-like filaments and other structures from flagellin. J. Mol. Biol. 9:168–185.

Lesman, M. C., A. S. Spirin, L. P. Gavrilova, and M. F. Golov. 1966. Studies on the structure of ribosomes. J. Mol. Biol. 15:268–281.

Sonneborn, T. M. 1963. Does preformed cell structure play an essential role in cell heredity? *In* J. M. Allen [ed.] The nature of biological diversity, pp. 165–221. McGraw Hill, New York.

Stern, C. 1968. Genetic mosaics and other essays. Harvard University Press, Cambridge, Mass.

Trinkaus, J. P. 1969. Cells into organs. The forces that shape the embryo. Prentice-Hall, Englewood Cliffs.

Ursprung, H. 1967. The formation of patterns in development. *In* M. Locke [ed.] Major problems in developmental biology, pp. 177–216. Academic Press, New York.

# Index